# SITE SURVEY AND EVALUATION ASPECTS FOR NUCLEAR INSTALLATIONS WITH EMPHASIS ON A GRADED APPROACH

The following States are Members of the International Atomic Energy Agency:

AFGHANISTAN
ALBANIA
ALGERIA
ANGOLA
ANTIGUA AND BARBUDA
ARGENTINA
ARMENIA
AUSTRALIA
AUSTRIA
AZERBAIJAN
BAHAMAS, THE
BAHRAIN
BANGLADESH
BARBADOS
BELARUS
BELGIUM
BELIZE
BENIN
BOLIVIA, PLURINATIONAL STATE OF
BOSNIA AND HERZEGOVINA
BOTSWANA
BRAZIL
BRUNEI DARUSSALAM
BULGARIA
BURKINA FASO
BURUNDI
CABO VERDE
CAMBODIA
CAMEROON
CANADA
CENTRAL AFRICAN REPUBLIC
CHAD
CHILE
CHINA
COLOMBIA
COMOROS
CONGO
COOK ISLANDS
COSTA RICA
CÔTE D'IVOIRE
CROATIA
CUBA
CYPRUS
CZECH REPUBLIC
DEMOCRATIC REPUBLIC OF THE CONGO
DENMARK
DJIBOUTI
DOMINICA
DOMINICAN REPUBLIC
ECUADOR
EGYPT
EL SALVADOR
ERITREA
ESTONIA
ESWATINI
ETHIOPIA
FIJI
FINLAND
FRANCE
GABON
GAMBIA, THE
GEORGIA
GERMANY
GHANA
GREECE
GRENADA
GUATEMALA
GUINEA
GUYANA
HAITI
HOLY SEE
HONDURAS
HUNGARY
ICELAND
INDIA
INDONESIA
IRAN, ISLAMIC REPUBLIC OF
IRAQ
IRELAND
ISRAEL
ITALY
JAMAICA
JAPAN
JORDAN
KAZAKHSTAN
KENYA
KOREA, REPUBLIC OF
KUWAIT
KYRGYZSTAN
LAO PEOPLE'S DEMOCRATIC REPUBLIC
LATVIA
LEBANON
LESOTHO
LIBERIA
LIBYA
LIECHTENSTEIN
LITHUANIA
LUXEMBOURG
MADAGASCAR
MALAWI
MALAYSIA
MALI
MALTA
MARSHALL ISLANDS
MAURITANIA
MAURITIUS
MEXICO
MONACO
MONGOLIA
MONTENEGRO
MOROCCO
MOZAMBIQUE
MYANMAR
NAMIBIA
NEPAL
NETHERLANDS, KINGDOM OF THE
NEW ZEALAND
NICARAGUA
NIGER
NIGERIA
NORTH MACEDONIA
NORWAY
OMAN
PAKISTAN
PALAU
PANAMA
PAPUA NEW GUINEA
PARAGUAY
PERU
PHILIPPINES
POLAND
PORTUGAL
QATAR
REPUBLIC OF MOLDOVA
ROMANIA
RUSSIAN FEDERATION
RWANDA
SAINT KITTS AND NEVIS
SAINT LUCIA
SAINT VINCENT AND THE GRENADINES
SAMOA
SAN MARINO
SAUDI ARABIA
SENEGAL
SERBIA
SEYCHELLES
SIERRA LEONE
SINGAPORE
SLOVAKIA
SLOVENIA
SOMALIA
SOUTH AFRICA
SPAIN
SRI LANKA
SUDAN
SWEDEN
SWITZERLAND
SYRIAN ARAB REPUBLIC
TAJIKISTAN
THAILAND
TOGO
TONGA
TRINIDAD AND TOBAGO
TUNISIA
TÜRKİYE
TURKMENISTAN
UGANDA
UKRAINE
UNITED ARAB EMIRATES
UNITED KINGDOM OF GREAT BRITAIN AND NORTHERN IRELAND
UNITED REPUBLIC OF TANZANIA
UNITED STATES OF AMERICA
URUGUAY
UZBEKISTAN
VANUATU
VENEZUELA, BOLIVARIAN REPUBLIC OF
VIET NAM
YEMEN
ZAMBIA
ZIMBABWE

The Agency's Statute was approved on 23 October 1956 by the Conference on the Statute of the IAEA held at United Nations Headquarters, New York; it entered into force on 29 July 1957. The Headquarters of the Agency are situated in Vienna. Its principal objective is "to accelerate and enlarge the contribution of atomic energy to peace, health and prosperity throughout the world".

IAEA-TECDOC-2101

# SITE SURVEY AND EVALUATION ASPECTS FOR NUCLEAR INSTALLATIONS WITH EMPHASIS ON A GRADED APPROACH

## GENERIC ROADMAP PROJECT — NUCLEAR SAFETY INFRASTRUCTURE FOR A FIRST NUCLEAR REACTOR: A HANDBOOK

INTERNATIONAL ATOMIC ENERGY AGENCY
VIENNA, 2025

Publishing Section
International Atomic Energy Agency
Vienna International Centre
PO Box 100
1400 Vienna, Austria
tel.: +43 1 2600 22529 or 22530
email: sales.publications@iaea.org
www.iaea.org/publications



For further information on this publication, please contact:

External Events Safety Section
International Atomic Energy Agency
Vienna International Centre
PO Box 100
1400 Vienna, Austria
Email: Official.Mail@iaea.org




Printed by the IAEA in Austria
October 2025
https://doi.org/10.61092/iaea.o4t2-2g3s

**IAEA Library Cataloguing in Publication Data**

Names: International Atomic Energy Agency.
Title: Site survey and evaluation aspects for nuclear installations with emphasis on a graded approach / International Atomic Energy Agency.
Description: Vienna : International Atomic Energy Agency, 2025. | Series: IAEA-TECDOC ISSN 1011-4289 ; no. 2101 | Includes bibliographical references.
Identifiers: IAEAL 25-01790 | ISBN 978-92-0-119025-3 (paperback : alk. paper) | ISBN 978-92-0-119125-0 (pdf)
Subjects: LCSH: Nuclear reactors — Design and construction. | Nuclear reactors — Safety measures. | Nuclear reactors — Technological innovations. | Nuclear reactors — Risk assessment. | Nuclear power plants.

## FOREWORD

In 2020, the IAEA launched the Generic RoadMap Project to support Member States in establishing appropriate safety infrastructure and improving the safety of nuclear installations. This project focuses mainly on providing technical support to countries embarking on nuclear installation projects, assisting them in developing the nuclear safety infrastructure for licensing their first nuclear power plant. It also supports countries planning to expand their existing nuclear power programme through the deployment of advanced nuclear power plants, including small modular reactors.

The site survey and site evaluation processes for a new nuclear installation are typically divided into five stages: site survey, site selection and site characterization, and the pre-operational and operational stages. Several important outputs are needed from the site survey and evaluation processes to support decisions affecting both safety and non-safety aspects of nuclear installations. When conducting site surveys and evaluations for advanced nuclear power plants, it is necessary to consider the risk associated with these installations by accounting for the reactors' type, size and intrinsic characteristics and safety features.

This publication provides practical information on site selection and characterization activities for nuclear installations, with particular emphasis on applying a graded approach to site evaluation for advanced nuclear power plants. It details various activities carried out at each stage of the site survey and site evaluation processes, and discusses the data and site attributes required at the different stages, with a focus on optimizing data collection efforts in accordance with the level of detail needed for advanced nuclear power plants.

This publication describes site characterization for advanced nuclear power plants on the basis of the radiological hazard categorization of these installations. First, a methodology for radiological hazard categorization of nuclear installations is described, which is based on a consequence analysis and takes into account the robust design and engineered safety features of advanced nuclear reactors. This is followed by a discussion on applying a graded approach to site investigations and characterization, with a focus on seismic and geotechnical hazard assessments, which are among the most resource intensive activities in the characterization of a nuclear installation site. A risk informed, performance based approach using a design level probabilistic safety assessment is also presented as a method for applying a graded approach to site evaluation for a nuclear installation.

This publication provides information to support compliance with the requirements established in IAEA Safety Standards Series No. SSR-1, Site Evaluation for Nuclear Installations, and the recommendations provided in IAEA Safety Standards Series No. SSG-16 (Rev. 1), Establishing the Safety Infrastructure for a Nuclear Power Programme. It also supports the recommendations provided in IAEA Safety Standards Series Nos SSG-35, SSG-79, SSG-18, SSG-9 (Rev. 1), SSG-93 and SSG-92 (these two publications are currently in preparation), and SSG-21, in relation to site selection, site evaluation and assessment of various natural and human induced external hazards.

The IAEA is grateful to all the experts who participated in the drafting and review of this publication, in particular F. Beltran (Spain). The IAEA officers responsible for this publication were M. Mahmood and Z. Gülerce of the Division of Nuclear Installation Safety.

## CONTENTS

# 1. INTRODUCTION

## 1.1. BACKGROUND

The IAEA supports States in establishing the appropriate safety infrastructure and enhancing the safety of nuclear installations throughout siting, design, construction and operation. This aim is achieved through the development of safety standards for nuclear installations and the provision of various services, such as peer reviews, education and training measures, seminars, workshops and lectures to support the effective application and implementation of the safety standards.

The IAEA safety standards are used, at the discretion of each Member State, as a reference to develop, update, and ensure nuclear installation safety. The IAEA General Conference continuously urges Member States to consider these safety standards in their nuclear power programme. In particular, the IAEA General Conference asks to the Secretariat to support embarking countries in the implementation of the safety standards during the development of a nuclear power programme.

The IAEA launched the project 'Generic Roadmap – Nuclear Safety Infrastructure for Licensing the First Nuclear Power Plant (NPP)' in 2019, which focuses on embarking countries and has the aim of supporting them in the development of a comprehensive nuclear safety infrastructure.

In the last decade an increasing number of States have considered incorporating nuclear power into their national energy policy. Currently, around 30 countries are considering or embarking on nuclear power generation, with IAEA supporting proper implementation of safety standards. Implementing a nuclear power programme necessitates meticulous planning, thorough preparation and significant investments of both time and resources. However, embarking countries likely have little experience in building, operating or regulating nuclear installations. Additionally, developing workers for the key organizations involved in the different phases of implementing a nuclear power programme is a common challenge for all countries embarking on this endeavour.

The life cycle of a NPP is divided into distinct stages, each featuring some key nuclear safety milestones. To achieve the different milestones, detailed attention needs to be given to processes of the site survey and site evaluation, which are closely linked to developing nuclear safety infrastructure. This is crucial for the systematic implementation of the first NPP in an embarking country, with specific roles for the government, regulatory body and the operating organization.

The IAEA is providing guidance related to evaluating both natural and human induced hazards, and subsequent design to meet the fundamental safety objective set forth in IAEA Safety Standards Series No. SF-1, Fundamental Safety Principles [1]. The fundamental safety objective is met through the use of standards comprising requirements established in IAEA Safety Standards Series No. SSR-1, Site Evaluation for Nuclear Installations [2] and recommendations provided in associated Safety Guides [3–9].

The Generic Roadmap Project [10] foresees the realization of prerequisites, publications and training modules to provide an easily understandable, well-structured and systematic approach to implementing the recommendations provided in IAEA Safety Standards Series No. SSG-16 (Rev.1) [11], as well as in other related IAEA safety standards. This TECDOC is one of the publications, specifically focusing on site survey and site evaluation for nuclear installations.

This publication is intended to provide practical guidance and support to States planning to extend their nuclear power programmes by deployment of advanced NPPs and to embarking countries who are planning their first nuclear power plant, including advanced NPP technology. This includes methodologies and examples on data collection, site investigations, site characterization and assessment of hazards to be carried out at various stages during the site survey and site evaluation processes for nuclear installations, with a focus on advanced NPPs. In particular, the application of a graded approach to the most resource-consuming activities of site investigation and hazards assessment are described.

## 1.2. OBJECTIVE

The objective of this publication is to provide practical and easily understandable guidance on how to select and evaluate a site for a nuclear installation. It includes a systematic and comprehensive approach that can be applied in a feasible and sustainable way given the multiple challenges and priorities of a particular nuclear installation. The activities of site survey and site evaluation described in this publication especially focus on (a) applying a graded approach for advanced reactors considering their intrinsic characteristics and passive safety features of the installation, and (b) optimizing resources devoted to the safety of the installation.

This publication provides guidance on how to apply a graded approach to site investigations and site hazard assessments based on the radiological consequences of potential releases from a particular nuclear installation, allowing optimization of resources during different stages of site survey and site evaluation. In line with IAEA safety standards, this publication does not aim to relax safety requirements, but rather to support their application through a graded approach. Examples of such an approach are provided.

The target audience of this publication includes operating organizations, regulatory bodies, research institutes, vendors and technical support organizations involved in the field of nuclear installation safety.

## 1.3. SCOPE

This publication discusses site survey and site evaluation within the context of nuclear safety infrastructure phases and the detailed activities for characterizing a site, focusing specifically on applying a graded approach, commensurate with the intrinsic characteristics and passive safety features of the particular nuclear installation. This includes NPPs, research reactors, advanced NPPs (including small modular reactors (SMRs)) and other nuclear installations. In addition, this publication describes the roles of the government, regulatory body and operating organization during site survey and site evaluation.

## 1.4. STRUCTURE

Section 2 provides detailed information on site survey and site evaluation within the context of nuclear safety infrastructure milestones and the framework of IAEA safety standards, including the role of the government, regulatory body and operating organization. Section 3 briefly describes the data and attributes needed at various stages of the site survey and site evaluation processes, with a discussion on minimizing the data needed for the site evaluation of nuclear installations categorized as intermediate or low hazard, taking into account the design of advanced safety features. Section 4 describes the different steps of applying a graded approach based on consequence analysis for the characterization of a nuclear installation site. In addition, it addresses the use of probabilistic safety assessment (PSA) to support the graded approach and discusses relevant regulatory considerations for review and assessment. Section 5 provides general guidelines for application of the management system to site survey and site evaluation activities. Final remarks are provided in Section 6.

Appendix I introduces a method for applying a graded approach to a probabilistic seismic hazard analysis (PSHA) for a site hosting an intermediate or low radiological hazard installation, located in a stable continental region. Appendix II introduces a performance-based approach for applying a graded approach to geotechnical investigations at a nuclear installation site.

Six annexes are organized to provide details on necessary data collection for site survey and evaluation process and some examples of site evaluation activities conducted in some Member States.

# 2. DESCRIPTION OF SITE SURVEY AND SITE EVALUATION PROCESS

This Section describes site survey and site evaluation processes for a nuclear installation. The role and function of the government, regulatory body and operating organization in the development of a nuclear power programme is also described. Additional considerations — including vendor's input, public involvement and some key aspects of advanced NPPs that are important to considered in site survey and site evaluation process — are also discussed.

## 2.1. GENERAL

The site survey and site evaluation processes described in IAEA Safety Standards Series No. SSG-35, Site Survey and Site Selection for Nuclear Installations [3] are closely linked with the IAEA milestones approach for developing the nuclear safety infrastructure needed to systematically implement the first NPP in a country. Figure 1 is reproduced from SSG-16 (Rev. 1) [11] and shows all the steps from the site survey to the commissioning stage, covering preparation and evaluation of a site evaluation report, site approval and the preparation and evaluation of a preliminary safety analysis report (PSAR) for obtaining the site construction licence from the regulatory body. The steps continue until the final safety analysis report (FSAR) is submitted to obtain authorization for commissioning of the nuclear installation.

It is important to note from Fig. 1 that the site survey and site selection need to be initiated in a timely manner and might become the critical path activity. Methods of streamlining the activities of site survey and site evaluation processes and supporting the application for a construction licence with sufficient and reliable information in a timely manner are also discussed in this section.

The time frames shown in Fig. 1 are suggested for large NPP projects; therefore, optimization of this timescale may be possible, especially for the deployment of advanced NPPs.

The siting of a nuclear installation consists of five stages, each designed to achieve specific milestones, as briefly outlined below [3, 11]:

(a) Site survey stage: Identifying potential regions, followed by potential sites and candidate sites through appropriate screening and comparison;
(b) Site selection stage: Evaluating and selecting final site(s) by ranking the candidate sites;
(c) Site characterization stage: Confirming site acceptability and completing site characterization, as well as deriving the site related design basis;
(d) Pre-operational stage: Performing work related to confirmatory and monitoring activities;
(e) Operational stage: Ongoing confirmatory and monitoring work, along with re-evaluation during periodic safety reviews.

Note that siting comprises both site survey and site selection and is usually a non-regulated activity in many States. Site evaluation starts after the site selection stage. Details of these stages are provided in Section 2.3. Figure 2 shows a graphical presentation of these stages [11].

Site survey and site evaluation activities to meet different milestones are as follows:

- One of the primary objectives of Milestone 1 during phase 1 is to conduct the site survey and identify a suitable candidate site (or sites); the nuclear power programme cannot move forward without a candidate site;
- During phase 2, a site is selected, evaluated and approved by the regulatory body. Design basis parameters are established, bids are prepared for invitation from potential nuclear suppliers, and the project is ready to invite bids in order to meet Milestone 2;
- During phase 3, the major activities are related to site preparation and construction of the NPP, including preparation and evaluation of the PSAR to obtain a construction licence, and preparation and evaluation of the FSAR to obtain an authorization for commissioning.

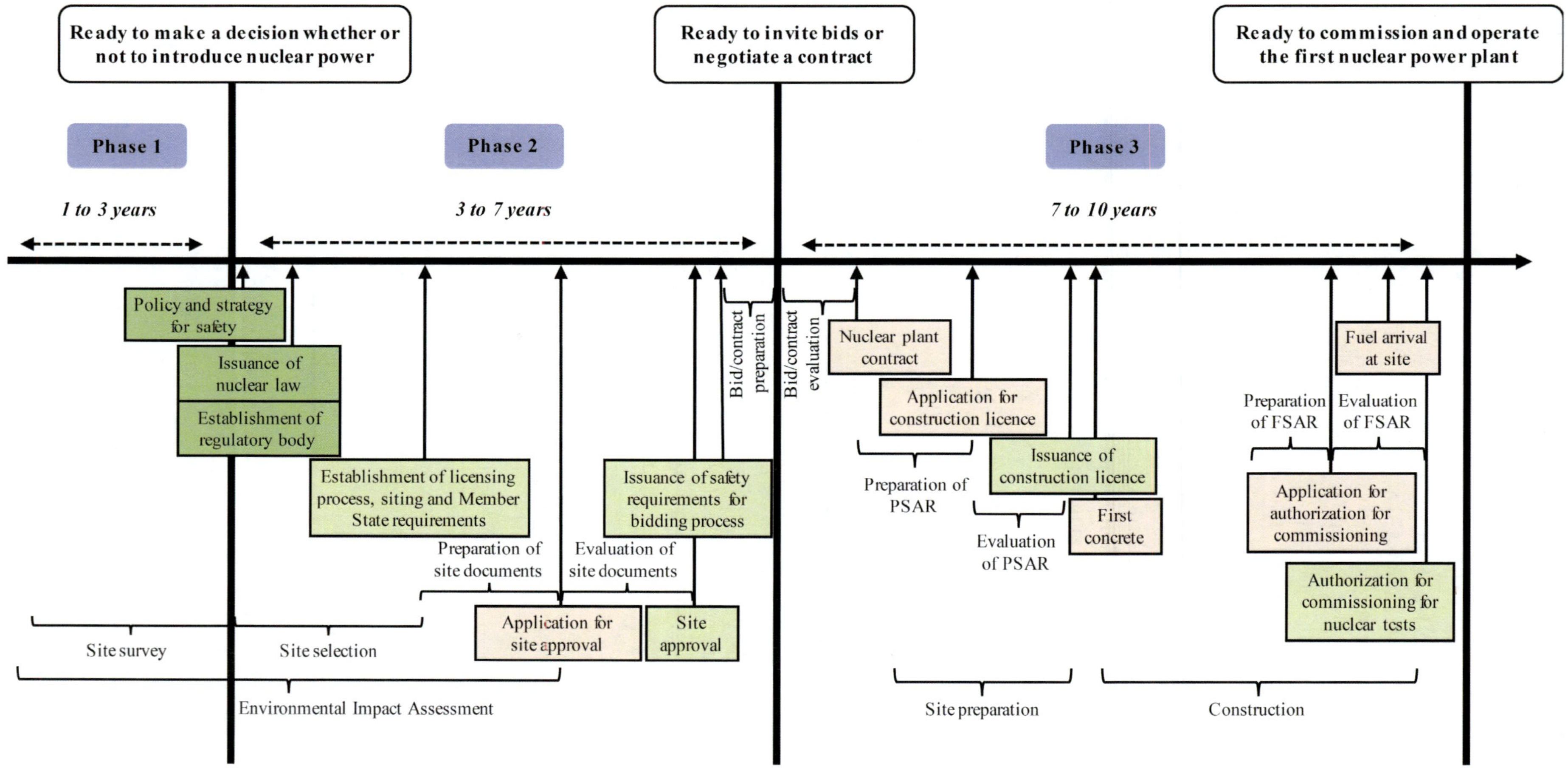


*FIG. 1. Indicative time frame and some important milestones for the development of a nuclear safety infrastructure (reproduced from Ref. [11]).*

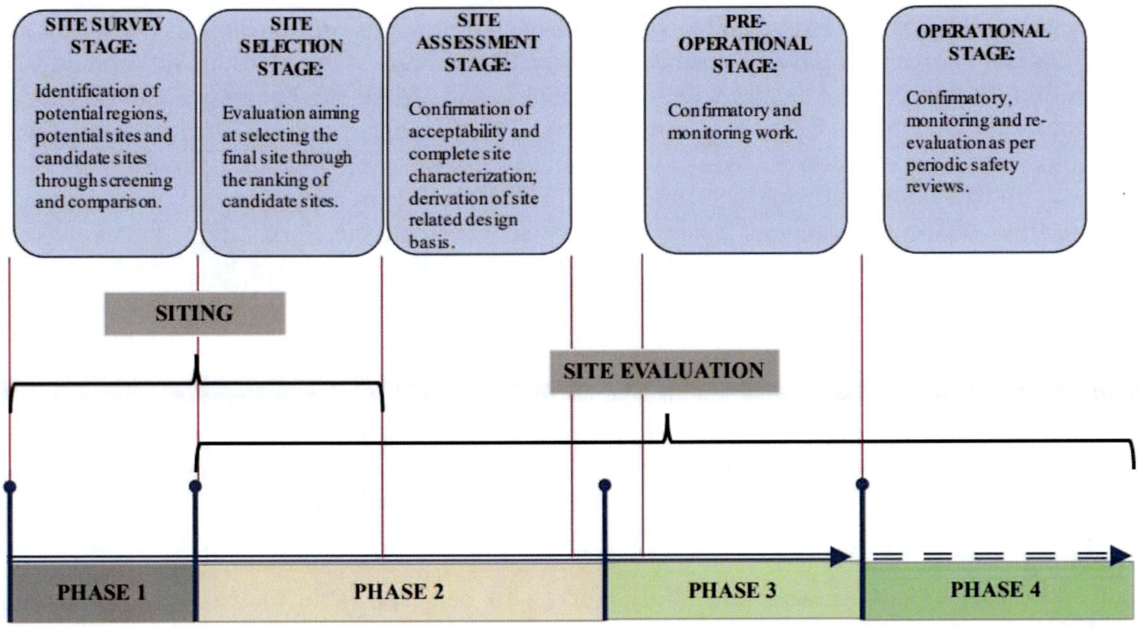


*FIG. 2. Consecutive stages of site survey and site evaluation (reproduced from Ref. [11]).*

Note that environmental aspects (i.e. environmental impact assessment (EIA), although considered during the stages of site survey and selection, are addressed through a parallel process at the site characterization stage. Different jurisdictions are usually involved, and interfaces must be properly defined. Similarly, the spatial planning for industrial or infrastructure projects is often regulated by national procedures, with their own requirements and milestones.

## 2.2. ROLES AND FUNCTIONS OF THE GOVERNMENT, REGULATORY BODY AND OPERATING ORGANIZATION

The government, regulatory body and operating organization have different roles and functions during each phase of developing a nuclear power project. The role of government is at a maximum during phase 1, gradually reduces during phase 2, and is at a minimum during phase 3. The role of the regulatory body starts with phase 2, gradually increases to a maximum during this phase and is maintained at maximum during phase 3. The role of the operating organization starts at phase 2 and gradually increases to a maximum until the end of phase 3. Siting and site evaluation activities vary during these phases.

A nuclear power programme is one of the major national projects that demands meticulous planning, extensive preparation and substantial investment of time, manpower and finances. Although not exceptional in this regard, nuclear power is thought to be distinct from other technologies due to the safety concerns raised by the ownership and handling of nuclear material as well as the ongoing commitment to ensuring safety once the decision has been made to start a nuclear power programme. A nuclear power programme is typically decided or chosen by the government due to the broad spectrum of issues that need to be considered, as well as the implications and length of the commitments involved. Both the government's strategy and policy decisions must consider and reflect the prime importance of safety [11].

The roles of the government, regulatory body and the operating organization during different phases in connection with the site survey and site evaluation process are as follows:

### 2.2.1. Phase 1

During phase 1, the government has a major role to play. During this phase, the country's sole nuclear power organization might be the Nuclear Energy Programme Implementing Organization (NEPIO). The work must be managed with the same quality and effectiveness during this phase as during phase 2, when the regulatory body and operating organization will develop integrated management systems. In order to properly manage and/or oversee the siting activities until integrated management systems are developed, the NEPIO has to establish a process that can effectively perform the regulatory body's functions in the course of these preliminary siting activities. These are needed to avoid duplication of effort and any project delays, as well as to provide the regulatory body, once it is formed, with the assurance that all previous work has been appropriately performed and assessed. While the organizations involved in such operations may still be emerging, a significant amount of work connected to site evaluation activities is carried out during phase 2. Phase 3 entails the formal confirmation of the design bases relating to the site, the creation of relevant licensing and approval procedures set by the regulatory body, continued monitoring of the site prior to operation, and the completion of all relevant licensing and approval requirements. The site will be monitored continuously throughout operations to ensure that it satisfies the design objective [11].

An effective regulatory body needs to be competent, well-resourced and strongly supported by the government. These are necessary for the national nuclear power programme's long term success, as well as the confidence of the international community and the public. Following the plans and suggestions developed by the NEPIO at the conclusion of phase 1, the regulatory framework will be established during phase 2.

At the beginning of the site characterization stage, the owner/operating organization may have already been established in embarking countries, or the NEPIO may still be responsible for carrying out programme activities [12].

Generally speaking, the thorough characterization of the chosen site(s) for the NPP should begin with clearly defined regulatory requirements in accordance with the goals to be met at this stage. One of the inputs that needs to be provided at the start of the site characterization includes these regulatory requirements.

At the pre-operational stage, the licensee (i.e. the owner/operating organization) must be fully established and prepared for operation, with an implemented integrated management system that defines its roles, responsibilities and functions [12].

SSG-16 (Rev. 1) [11] states:

> "Action 110. The operating organization should update the radiological environmental impact assessment for the site selected, as appropriate.
>
> "Action 111. The regulatory body should review and assess the radiological environmental impact assessment for the site selected, as appropriate.
>
> "Action 112. The operating organization should commence a radiological environmental monitoring programme."

### 2.2.2. Phase 2

During this phase, preparatory work on the safety infrastructure for constructing a nuclear power plant is undertaken following a policy decision to pursue nuclear power. The regulatory body is formed, establishes the licensing process, prepares the relevant safety regulations, evaluates the site evaluation report for the acceptability of the selected site(s) and issues the site(s) permit.

The operating organization selects the preferred candidate site, establishes its acceptability, completes detailed investigations, characterizes the site and prepares the site evaluation report, including the design basis parameters related to the site. This report is then submitted to the regulatory body for approval of the site. Additionally, the operating organization uses the site information, including regulatory controls, to prepare the bid specifications for inviting bids or negotiating a contract for the first nuclear power plant.

SSG-16 (Rev. 1) [11] states:

"Action 161. The regulatory body should establish specific safety requirements for site evaluation, including requirements for the process of authorizing the site selected, in compliance with the relevant IAEA safety standards.

"Action 162. The operating organization should complete the investigations relating to the acceptability of the candidate sites and should select the preferred candidate site for the first nuclear power plant, making use of specific data, information and studies, as well as assessments conducted on the fullest possible temporal and spatial scales of investigation.

"Action 163. The operating organization should prepare the site evaluation report and should submit it to the regulatory body on the basis of a full assessment of the site selected and including the confirmation of site acceptability and the characterization of the site for the definition of the site related design basis parameters.

"Action 164. The regulatory body should review and assess the site evaluation report and should make a decision with regard to the acceptability of the site selected and the site related design basis.

"Action 165. The operating organization should use all the appropriate information relevant to safety and to regulatory control that relates to or is derived from the site characterization to prepare the bid specifications or to negotiate a contract for the first nuclear power plant.

"Action 166. The operating organization should start to evaluate and modify the site and the radiological environmental monitoring programme, as necessary, after the site evaluation report has been approved."

### 2.2.3. Phase 3

The operating organization evaluates the bids and signs the contract for the design, construction and commissioning of the nuclear power plant [11]. It also updates the site evaluation report to reflect the specifics of the selected nuclear power plant and incorporates data collected during the pre-operational stage. It also implements the site safety improvements, as a result of external hazard assessments and continues to implement the site monitoring programme.

SSG-16 (Rev. 1) [11] states:

"Action 167. The operating organization should prepare the chapter on site evaluation in the safety analysis report and then update it, taking into account the specificities of the nuclear power plant selected and the data and information gathered during the pre-operational stage.

"Action 168. The operating organization should implement necessary safety improvements to the site, as site protection measures determined as a result of the external hazard assessment.

"Action 169. The operating organization should continue to implement the environmental programme and the site monitoring programme."

More detailed recommendations on actions to be completed by the government, regulatory body and operating organization at each phase of site survey and site evaluation process are provided in SSG-16 (Rev. 1) [11].

## 2.3. STAGES OF SITE SURVEY AND SITE EVALUATION

The site survey and site evaluation processes involve five distinct stages, which begin at the outset of the nuclear power programme and continue throughout the entire life cycle of the nuclear installation. These stages are discussed in following subsections.

### 2.3.1. Site survey stage

The site survey stage is the initial phase of the siting process and is extensively described in SSG-35 [3], where emphasis is put on the safety requirements established in SSR-1 [2] and NG-T-3.7 (Rev.1) [12], which focuses on developing national infrastructure for nuclear power. Fig. 3 shows the general siting process in a flow chart, including site survey and selection for a site of a nuclear installation [3].

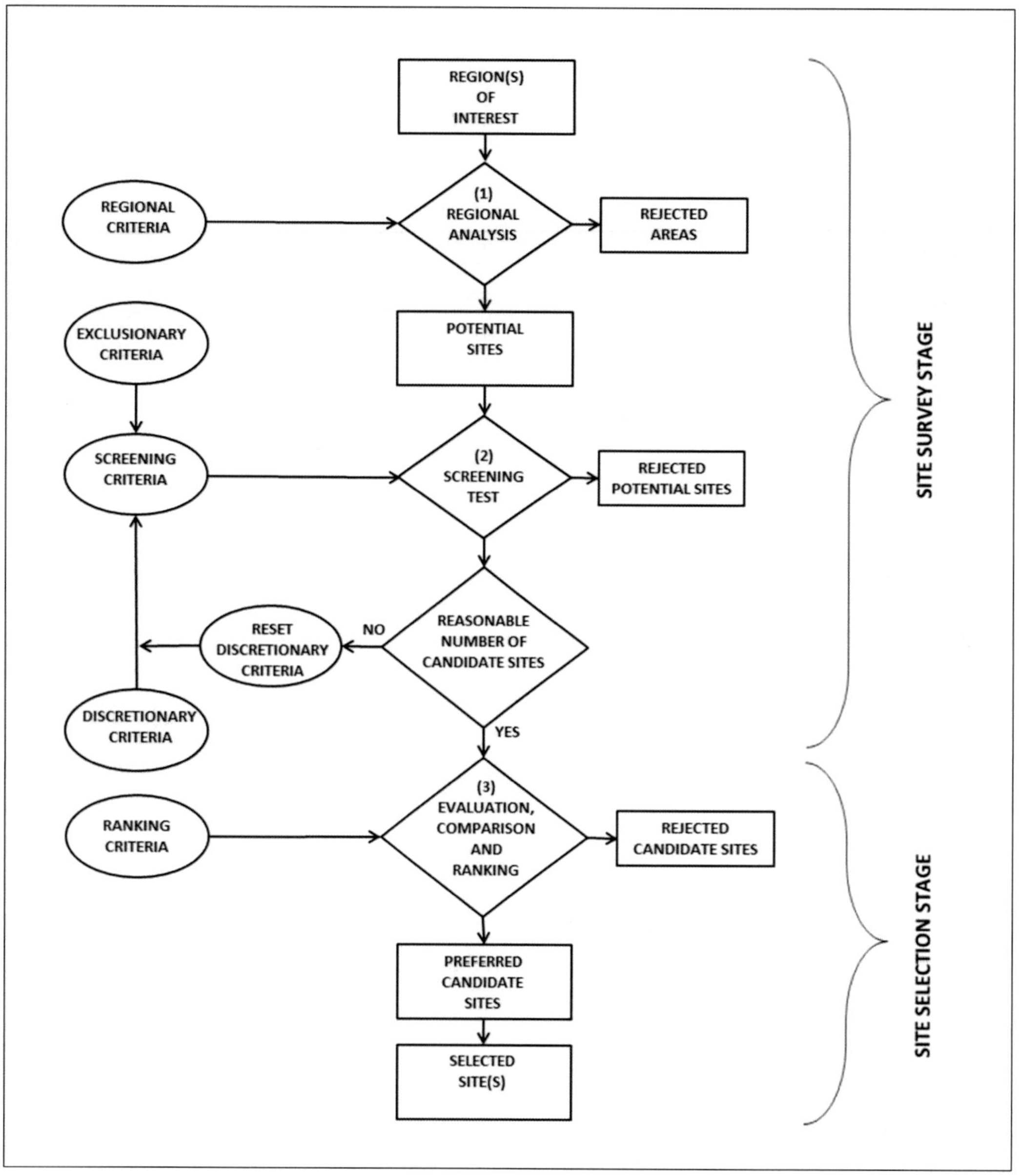


*FIG. 3. Flow chart for the siting process for a nuclear installation (reproduced from Ref. [3]).*

As stated in Section 2.1, the site survey is often a non-regulated activity, driven by the nuclear project proponent. However, if the regulatory body is not yet established during the siting process, government institutions should be consulted and concur on siting decisions if a non-government entity is leading the siting effort.

#### *2.3.1.1. Regions of interest*

The site survey starts with the regional analysis, during which regions of interest are examined to identify potential regions and, subsequent potential sites within those regions. During the initial site survey stage, the identification of potential sites and candidate sites is initiated once the project objectives have been clearly defined (i.e. power and type of the plant to be built, number of units, target users, if any) and the development phase or feasibility study is launched. A key element is the definition of the geographical scope of the site survey, through the so-called 'regions of interest.' Practical experience shows that, even for a State-wide survey, the identification of regions of interest is a necessary first step that will help focus resources and avoid unnecessary effort.

The regional analysis is based on well-established regional criteria, which are as stated in para. 3.7 of SSG-35 [3] "generally related to national domestic policy, national economic, national and international environmental protection policy or other related policies of the State." However, it is important to consider other siting criteria before embarking on a site survey as it will greatly help to quickly eliminate unsuitable areas and thereby minimize the total time and resources devoted to survey work. The exclusion criteria applied during the next step are discussed in Section 2.3.3. Moreover, there are country-specific exclusionary areas (e.g. the River Nile in Egypt), and some regions in some Member States must be excluded due to security issues. Regions with technical constraints, deprived of cooling water or lacking infrastructure are expected to be excluded for large power plants feeding the electrical network. However, they may be included if the project includes the option of off-grid advanced NPPs, potentially equipped with air-cooled condensers and supported by safety analysis.

While newcomer countries will typically consider a vast territory as starting point, a more focused approach may be applied if the goal is to repower existing assets, such as coal fired power plants or industrial sites using fossil fuels. In such a case, the regions of interest may be substituted by a discrete list of specific locations, i.e. existing (brownfield) sites, rather than large stretches of land. Care is needed not to discard potential areas that might present a high level of safety combined with strong socioeconomic benefit.

#### *2.3.1.2. Data collection for site survey*

As recommended in SSG-35 [3], the site survey stage will include the data collection effort needed to feed the desk-top studies supporting the survey process. The work is to be conducted under a robust quality management system that will ensure that the data collected are relevant, up-to-date and reliable [3].

Paragraph 7.1 of SSG-35 [3] states:

> "The siting process should be addressed in the overall management system for the nuclear installation project. The management system for siting should be established at the earliest possible time consistent with its implementation in the conduct of activities for the site survey and site selection stages of the nuclear installation."

The database to be constructed is a geographically indexed database (geographical information system, or GIS) that will be progressively enriched through the siting and site evaluation processes.

The GIS database contains multiple layers of information according to the technical disciplines, such as: geological capable fault maps, population density, environmental restrictions, cultural heritage, flooding zones, hazardous facilities, land use, meteorological data.

The Appendix in SSG-35 [3] provides recommendations on the database for the siting process (site survey and selection stages).

#### *2.3.1.3. Screening criteria*

Potential sites are identified after data collection, as outlined in the subsections 2.3.1.1 and 2.3.1.2, followed by a screening process. It is important to mention that the data collected during the site survey stage, as stated in para. 5.1. of SSG-35 [3] are based on "information and data collected principally from existing sources such as available records, satellite images, topographic sheets and information available from relevant national and local authorities and other institutions." However, if these data are insufficient or deemed of inadequate quality during the regional analysis, an effort to assemble more data, including limited fieldwork or investigations, are initiated so that the results are available during the site selection stage.

Screening criteria are used to choose candidate sites. These are defined upfront, properly documented and comprise of two types:

— Exclusionary criteria for which no practical mitigation measure can be found, such as environmental exclusion zones, cultural heritage locations, areas where it is not feasible to implement emergency plans, geological faults that could cause displacement at or near the surface, other areas that are prone to massive soil instability, areas that are exposed to flooding, to severe human induced hazards. In most cases, a buffer zone or safety distance may have to be defined around those exclusion zones or hazardous features.
— Discretionary or avoidance criteria for locations that are inconvenient or entail too much project risk or cost, even though an engineered solution can be found to ensure site safety. These are typically related to conditions such as the distance to cooling water, unfavourable topography or bathymetry, poor soil conditions, built-up or recreational areas, lack of infrastructure (e.g. grid, transport routes), high population density.

The exclusionary and discretionary screening criteria for safety related and non-safety related aspects listed in Annex I are reproduced from SSG-35 [3].

#### *2.3.1.4. Identification of candidate sites*

Candidate sites are identified after applying the screening process (using both exclusionary criteria and discretionary criteria) to the potential sites.

There is a certain level of adjustability to the discretionary criteria, so that the number of screened-in sites correspond to a manageable set of options for defining candidate sites. This is typically an iterative process, where criteria may be adjusted to be more or less stringent based on the desired number of candidate sites for further evaluation. During this iterative process for the identification of candidate sites, as stated in para. 3.15 of SSG-35 [3], "if candidate sites are distributed in two or more regions with different attributes, this would preclude the possibility of the elimination of all the candidate sites on the basis of common regional shortcomings."

A better alternative is to develop a set of evaluation criteria (corresponding to the above discretionary criteria) and to model them into the GIS software, so as to perform a multi-criteria analysis of the potential (not excluded) areas. Each criterion will yield a particular score, and the scores will then be combined using a weighted sum method. Alternatively, a minimum score for each criterion can be defined and the zones satisfying the minimum score on all criteria can be plotted.

The site survey stage is completed when a discrete list of candidate sites has been identified. The chosen candidate sites need to present a sufficient level of diversity. There needs to be reasonable assurance that these sites fulfil all safety requirements; this assurance is likely to be further enhanced during the site selection stage when additional data and/or limited fieldwork is undertaken. Final confirmation will only be obtained at the early site characterization stage when detailed studies are conducted to completely rule out the exclusion criteria.

### 2.3.2. Site selection stage

In the site selection stage, the identified candidate sites are evaluated, compared and ranked. The objective of this step is to assess the sites to ensure there are no features at or around the site areas that would prevent the construction and operation of the nuclear installation. Additionally, this stage involves comparing and ranking the sites based on their suitability and attractiveness as potential locations for the nuclear installation [3].

Ranking criteria provide bases for comparing candidate sites, enabling the selection of a list of preferred candidate sites. As stated in para. 3.20 of SSG-35 [3], "ranking criteria are generally developed by using considerations relating to discretionary criteria along with relevant non-safety-related considerations." Additional data for candidate sites is generally collected through fieldwork.

#### *2.3.2.1. Data collection for candidate sites*

Sufficient data must be obtained to support the comparison and ranking of the candidate sites. As much as possible, the data used for comparison need to be uniform in amount and quality across the regions or sites being compared. Comprehensive surveys of the relevant literature are necessary, but usually not sufficient, during the site selection stage. An initial work programme should be planned during the site survey stage in case the data on an external hazard for a potential site is insufficient. This programme includes targeted field investigations (e.g. subsoil conditions, flooding risk).

The appendix to SSG-35 [3] provides recommendations on constructing the database for the siting process (site survey and selection stages).

#### *2.3.2.2. Preliminary estimates of radiological impact and the emergency planning zone*

Prior to ranking of the candidate sites, a preliminary estimate is needed to understand the implications of site selection on the environment, for normal and accidental releases, and for the size of the emergency planning zone (EPZ) for accidental releases. Such information should be requested from the equipment vendors and must be included in the plant parameter envelope and supporting documentation (see Section 2.8.1).

#### *2.3.2.3. Scoring and ranking*

Once candidate sites are identified through the screening process presented in Section 2.3, a scoring and ranking system is used to create a list of preferred candidate sites. The process must be transparent and involve primary stakeholders, including public participation.

Multiple decision making methodologies exist; the suitability index method is often used and has been applied successfully in numerous site selection projects. It consists of the following steps:

(a) Definition of site selection attributes that will allow differentiating the candidate sites; these attributes include safety, environment, cost and socioeconomic impact.
(b) Definition of attribute groups that gather the different attributes under the above categories and are associated with a technical discipline (for example: geology).
(c) Definition of a scoring system in agreement with the owner/project developer and the main stakeholders. One scoring system typically applied is the ordinal scaling that will rank the attribute in an order of preference on the basis of a discrete or continuous scale.
(d) Definition of score scaling for each attribute by experts in the technical discipline. The score scaling can then be approved by the owner and main stakeholders.
(e) Allocation of weights to all the attributes, reflecting their relative importance. Initial weights are assigned with input from primary stakeholders via dedicated workshops. Weights are then adjusted by multi-disciplinary experts in consultation with the owner, to eliminate potential bias introduced through tunnel vision.
(f) Score evaluation for each attribute and each candidate site, using the score scaling previously defined. This evaluation is conducted by experts in the technical disciplines.
(g) A Suitability Index Score is obtained for each candidate site, through the formula given below. Candidate sites are ranked using the Suitability Index Score;

$$S_j = \sum_{i=1}^{n} \text{weight}_i \ \text{score}_i \tag{1}$$

Where:
$S_j$ is the Suitability Index Score of candidate site j;

Weight$_i$ and Score$_i$ are respectively the specific weight (in %) and evaluated score of attribute i for the candidate site j;

i is the considered attribute; $n$ is the total number of attributes.

(h) Presenting the results in a workshop, with stakeholder participation.

A good practice is to use the GIS based multi-criteria analysis tool mentioned in Section 2.3.1.4 to inform the scoring process, through colour mapping for those attributes that can be mapped.

As expressed earlier, it is important to have a strong technical basis for the scores. Subsoil conditions in particular can be substantiated by sufficient data.

#### *2.3.2.4. Preferred site(s)*

The scoring and ranking of the candidate sites will result in a shortlist of preferred candidate sites, from which a site may be selected for the nuclear installation [3]. At least one alternative site should be identified, in case the site characterization stage reveals unfavourable conditions (e.g. the presence of potentially capable fault in the site vicinity). The government or operating organization generally make the final decision on selection of site for the nuclear installation, consulting with all the relevant stakeholders.

A site selection or site suitability report is the recommended deliverable for this stage of the project.

### 2.3.3. Site characterization stage

The site characterization stage, which includes confirming site acceptance and deriving site related design basis parameters (also known as site assessment) begins after an NPP site is selected following site survey and selection stages.

#### *2.3.3.1. Site characterization programme*

The scope and extent of the site characterization programme must be defined, taking into account all applicable safety requirements as defined in SSR-1 [2], the relevant Safety Guides [3–9] and national regulatory requirements and guides. The typical documents to be prepared include site evaluation report, EIA and Chapter 2 of the Safety Analysis Report (PSAR, FSAR, permitting procedures). The list of site design basis parameters also needs to be defined, as an output of site characterization.

The site characterization aspects covering the IAEA safety requirements and recommendations along with the associated data to be collected are extensively described in Section 3. A good practice is to build a site characterization requirements matrix that provides a clear mapping between the requirements, the site characterization activities (including the standards to be applied) and the reporting of the results.

Paragraph 3.24 of SF-1 [1] states:

> "The resources devoted to safety by the licensee, and the scope and stringency of regulations and their application, have to be commensurate with the magnitude of the radiation risks and their amenability to control. Regulatory control may not be needed where this is not warranted by the magnitude of the radiation risks."

As a consequence, more time and resources need to be devoted to those hazards that contribute significantly to the risk to the public. The site characterization programme for a given site has therefore to include discussion on the potential significance of each risk contributor.

#### *2.3.3.2. Site characterization activities*

It is customary to group the activities according to the technical disciplines. Data collection includes site surveys and field investigations to obtain the relevant data, either directly or through laboratory tests. A permit to access the land will have to be obtained. Hazard assessments are performed to characterize each hazard in accordance with the requirements established in SSR-1 [2] and recommendations and guidance provided in:

- Seismic hazard assessment (SSG-9 Rev. 1 [6]);
- Geotechnical aspects and considerations for foundations (SSG-93 [7]);
- Volcanic hazards (SSG-21 [9]);
- Meteorological and flooding hazards (SSG-18 [5]);
- Hazards associated with human induced external events (SSG-79 [4]);
- Dispersion of radioactive material in air and water, population distribution, use of land and water and consideration of the feasibility of an emergency plan (SSG-92 [8]).

*2.3.3.3. Derivation of site specific design basis*

As a continuation of the previous section, the site specific design parameters are derived considering target annual probabilities of exceedance or using deterministic methods. SSG-67, Seismic Design for Nuclear Installations [13] and SSG-68, Design of Nuclear Installations Against External Events Excluding Earthquakes [14] provide recommendations on design for seismic and other hazards.

Design basis seismic hazard values are typically defined at or below a $10^{-4}$ annual probability of occurrence. Consideration has also to be given in the safety analysis to potential cliff edge effects and to confirm that adequate margin in the design of the plant is provided for protecting items that are important to safety. Hazards need to be quantified down to a probability level below which such cliff edge effects and their consequences can be considered to be 'practically eliminated.'

A comparison table may be prepared comparing the site specific design values to the generic design values for a standard design used by the vendor. This will allow readily an estimation of the design margins and a confirmation that Safety Principles 5 and 8 of SF-1 [1] are fulfilled, i.e.

> Principle 5: Optimization of protection: "Protection must be optimized to provide the highest level of safety that can reasonably be achieved."
>
> Principle 8: Prevention of accidents: "All practical efforts must be made to prevent and mitigate nuclear or radiation accidents."

*2.3.3.4. Reporting*

The preparation of a comprehensive site evaluation report along with specific requirements of a State will support the application for site approval by the regulatory body. The site evaluation report is based on topical reports of various site hazard evaluations.

A compliance matrix may be included in the site evaluation report, demonstrating compliance with the requirements established in SSR-1 [2] (with reference to topical reports chapters) or, by exception, justifying departures from these requirements.

Chapter 2 of the PSAR is also a cornerstone document in which site characteristics will be summarized and the plant design basis provided explicitly.

### 2.3.4. Pre-operational stage

The pre-operational stage for a nuclear installation includes activities that span from commencement of the construction to onset of operation such as design, site preparation, construction, assembly and commissioning. At the pre-operational stage, site specific activities with regard to confirmatory and monitoring work are undertaken. In order to conclude and refine the site characteristics assessment, studies and investigations started in the preceding stages are continued [3]. The collected site data provide a final evaluation of the simulation models applied to the final design. During the pre-operational stage, examples of major activities include implementing a monitoring programme based on the site characterization studies. This includes detailed geotechnical investigations carried out on the basis of final layout of structures and buildings, with consideration of safety-related structures, visual inspections/mapping of geological/geotechnical strata, and laying of geotechnical instrumentation (e.g. load cells, settlement meters, piezometers, inclinometers) below the safety and non-safety-related buildings and structures.

### 2.3.5. Operational stage

Site evaluation activities related to safety are conducted over the operating lifetime of the nuclear installation, primarily through site specific monitoring and periodic safety review. This includes, as stated in para. 2.5 of SSG-35 [3], "all confirmatory, monitoring and re-evaluation work conducted throughout the operational stage, and especially during periodic safety reviews of the installation." This work is generally reported in periodic safety review reports every ten years. The monitoring follows the established site monitoring programme and includes monitoring of seismic, geotechnical, meteorological, groundwater, radiological and flooding parameters.

## 2.4. ADDITIONAL CONSIDERATIONS

Some more considerations in relation to input from the vendor, involvement of the public, specific aspects of advanced NPPs, planning for the project, and regional analyses are described in following subsections.

### 2.4.1. Vendor's input

Technical dialogue between NEPIO, or the operating organization, with the vendors is of prime importance during different phases of a nuclear installation project. Most vendors have developed a generic design based on a set of bounding site parameters. Identifying and selecting sites that fit within the site parameter envelope will guarantee that the generic design can be applied, and this has important implications on the licensing process and cost of the project.

Site design parameters include (not an exhaustive list):

- Generic design ground motion response spectra;
- Maximum groundwater table elevation;
- Maximum flood level;
- Minimum soil properties in terms of soil bearing capacity and allowable settlement at the foundation level;
- Range of soil parameters (e.g. minimum shear wave velocity), considered in both the static and dynamic analysis of the building structures, bounding values of ambient air temperature, humidity, precipitation, wind speed, tornado, etc.;
- Bounding values of pressure waves from external human induced hazards.

In addition, most vendors have prepared interface data that characterize the facility's environmental impact, as follows:

- Minimum site footprint for the finished site and the construction site area;
- Minimum cooling water requirements;
- Chemical and radiological releases for different plant conditions;
- Noise;
- Visual impact.

Those data are usually compiled in the form of a plant parameters envelope encompassing the different reactor technologies considered for the project, in effect defining a surrogate plant whose characteristics can be used in early site permitting procedures. Examples can be found in Ref. [15].

To be effective, the site survey stage needs to proceed on the basis of a known subset of site and plant design parameters (e.g. minimum plant footprint, cooling water requirements, minimum soil properties, bounding seismic hazard and flood levels at $10^{-4}$ annual probability exceedance value, bounding EPZ size).

At the site selection stage, the preliminary characteristics of the candidate sites need to be checked against the vendor data; preliminary field investigations need to be conducted in support of this verification. It is also a good practice to share preliminary site layout information and results of preliminary surveys with the prospective vendors.

Site characterization activities also take into account vendor inputs such as the standard borehole pattern and requirements for geotechnical characterization.

### 2.4.2. Communication and public acceptance

Communication on the future nuclear plant project needs to take place early on. Involvement of the governmental, regional and municipal authorities, general public, potentially neighbouring States and, in particular, communities in the site vicinity, is to be planned as part of a well-defined stakeholder engagement policy.

As discussed in Sections 2.4.3 and 2.4.4, direct participation is assumed to take place in an open and transparent site selection process, to designate the preferred site(s).

Raising the public awareness about the safety fundamentals, residual risk, emergency planning, environmental impact during all project phases, and the social and economic impact of the nuclear plant project is to be proactively managed in parallel with the site evaluation processes.

### 2.4.3. Advanced nuclear power plants

Generally, the international guidelines and national regulations on the subject of siting of nuclear installations were prepared for large power plants, for which radiological consequences of accidents could affect a large territory. These constitute a comprehensive framework within which to carry out the limited adaptations that are deemed justified to reflect the specifics of advanced NPPs. The justification for those adaptations is discussed below. However, eventual compliance with regulations and international standards will always have to be demonstrated.

#### *2.4.3.1. Reduced size of the emergency planning zone*

It is often claimed that advanced NPPs are intrinsically safer, having a smaller core and relying on passive safety systems that require neither human intervention nor external or backup power to operate. Some advanced reactor vendors are demonstrating that the EPZ for their reactors can be limited to the site boundary. However, it is the responsibility of States whether to accept such claims from vendors.

The evaluation of the radiological consequences of accidents, and the justification for reduced EPZ size, is an important element of the licensing of advanced NPP technology, which will follow a parallel path to the siting activity.

At the site selection stage, a preliminary estimate of the maximum radiological impact and EPZ needs to be available. This will provide sufficient assurance that emergency planning measures, if necessary to assure the fifth level of defence-in-depth, are feasible given the population density and the presence of critical facilities (e.g. hospitals, schools) in the immediate site vicinity.

While IAEA safety standards do not prescribe population density limits within a certain perimeter, prudence suggests avoiding residential and built-up areas, for social acceptance reasons. The level of prudence to be applied needs to reflect the type of advanced NPP, i.e. micro-modular reactors versus grid-scale, multi-unit SMRs, and associated maximum dose calculations.

Conversely, in cases where there is no radiological impact beyond the site boundary, or an impact only over a very short range, sites near to population centres that offer a range of services and qualified workforce might be considered for advanced NPPs. Moreover, an advanced NPP near a population centre can more efficiently deliver power and heat, in particular district heating.

#### *2.4.3.2. Co-location with industrial facilities*

Electricity decarbonization objectives will make the case for co-location of advanced NPPs with industrial plants more frequent, or even a new paradigm. In such cases, an in-depth assessment of reciprocal hazards (hazard from the industrial facility on the safe operation of the reactor; nuclear related hazard from the reactor that could affect industrial plant's safe operation; combined emergency planning) must be conducted.

#### *2.4.3.3. Multiple sites*

If a fleet of SMRs are envisaged in a State, in order to achieve economy of scale and to respond to the broad decarbonization needs, a different approach to the individual site selection and characterization could be adopted, with all due caution and respect for the current regulatory framework. As a minimum, it would consist of conducting the site survey and evaluation activities as a single project using a consistent set of procedures and a repetitive process.

### 2.4.4. Planning

The timeline provided at the beginning of this section underlines the importance of initiating the siting activities early. While for many States it is not necessary to apply for a site approval specifically and distinctly from the construction licence application, the site characterization activities are necessary to provide a firm ground for contracting the project.

Below are activities that may dictate the critical path:

#### *2.4.4.1. Probabilistic seismic hazard analysis*

Derivation of the site specific seismic hazard and of the seismic design basis for the plant, in accordance with the recommendations provided in SSG-9 (Rev. 1), Seismic Hazards in Site Evaluation for Nuclear Installations [6] and national procedures, is a long process that has to be initiated right from the site selection stage, or even earlier. An optimized (simplified, accelerated) procedure for areas outside of active tectonic regions has been proposed and is discussed in Section 4.

#### *2.4.4.2. Environmental studies*

The EIA first requires the establishment of a solid environmental baseline describing the initial status of the selected site. This will involve detailed field surveys. The impact of the nuclear installation to be built then needs to be defined and assessed. If the technology and associated plant parameters have not yet been defined because of the contractual process, it will be necessary to construct a plant parameters envelope, as discussed in Section 2.7.1. This plant parameters envelope will form the basis for performing the EIA and preparing an application for review by the national environmental agency.

#### *2.4.4.3. Meteorological study*

The establishment of site specific parameters for the meteorological conditions, per current guidelines, requires a minimum of one year of monitoring [8]. This implies installing a fully equipped meteorological monitoring mast at the selected site, accumulating data over at least one year.

### 2.4.5. Use of regional analyses for multiple sites

In consideration of the time constraints highlighted above and based on experience from useful precedents in certain areas (e.g. regional seismic hazard maps), an optimized and uniform process is suggested to be applied at the regional level, encompassing a range of site locations, prior to site specific work. The regional scale characterization, where feasible (e.g. seismicity, meteorology), should precede the local scale characterization, thereby reducing the scope and duration of the latter. This will ensure that a consistent and cost-effective approach, as well as common data sources, will be used for the site attributes and characteristics that are not so dependent on the exact location of a particular site. This approach will be further beneficial as plant designs become more standardized, and plants can be sited based on more generic site parameters that may cover large areas. Plant designs could then be deployed with little or no site specific adaptation.

It therefore becomes an essential objective of a siting programme to verify that the site design parameters (which can be extracted from the plant parameters envelope prepared by vendors) are not exceeded at the regional level, encompassing multiple sites.

The definition of the regional level matches the 'regions of interest,' which is the starting point of a siting project as described in Section 2.3, within which the candidate sites are located.

The output of the regional analysis will be in the form of maps detailing the exclusion zones, the severity of natural and human-made hazards (hazard maps) and meteorological parameters. It will also contain topical reports that can be peer-reviewed, and a project database based on the recommendations provided in SSG-35 [3].

The local aspects of site evaluation (i.e. within site vicinity) are to be addressed in a second step that is site specific; the local scale may be defined by a range of minimum 5 km around the nuclear site, the latter being defined by its protected area boundary that encloses all radiologically sensitive elements of the plant.

The main elements and benefits of this approach are highlighted for the following two examples:

- Seismic hazard analysis: the possibility to initiate the regional and near-regional analyses in parallel with the site selection process; to construct the seismic catalogue containing the history of seismic activity in the region; to characterize the main sources of seismic activity; to choose ground motion predictive equations that are representative of the regional and subregional geological setting; to construct the logic trees. In this way only the site specific parameters (local site effects) and location dependent hazard computations are left for the local hazard quantification. In addition, conservative estimates of the ground motion parameters can be obtained early on and compared to generic design parameters to establish the preliminary safety margins.
- Meteorological analysis: identification of meteorological stations in the region that satisfy the quality requirements of the World Meteorological Organization; collection of meteorological records; statistical analysis to determine the variability of key parameters (e.g. average temperatures, extreme values of temperature, wind, precipitation) across the region; determination whether the data variability and the site configuration (local topography) warrant the installation of a meteorological monitoring mast for long term monitoring; estimation of climate change effects at the regional level; use of conservative envelopes to compare regional and site data to the vendor's site design parameters. It is anticipated that in many countries, meteorology will not be a discriminating factor between sites and that comfortable margins can be evidenced against the vendor generic design data, thereby smoothing and shortening the site characterization process.

# 3. DATA COLLECTION FOR SITE EVALUATION

## 3.1. GENERAL CONSIDERATIONS ON DATA COLLECTION

The safety of nuclear installations against natural and human induced external hazards can be ensured by adequately characterizing these hazards and implementing proven design solutions to deal with the effects of these hazards. To adequately characterize a site for a nuclear installation, data collection and use of a comprehensive database in hazard evaluations plays a significant role in obtaining realistic design bases parameters. IAEA safety standards [2–9] provide the requirements and recommendations for compiling a comprehensive database on relevant site characteristics needed during various site survey and site evaluation stages.

An initial assessment of the preliminary design bases for a nuclear installation can be made using available information on regional characteristics. This assessment gradually improves as more information on site characteristics is collected during the site survey stage, until the detailed studies are completed during site characterization stage.

The database needed for the site survey and site selection stages has been discussed in Section 2. Figure 3 shows the complete process of site survey and selection. The exclusion criteria described in Section 2.3.3 need to be carefully addressed during the site selection stage so that implementation of the nuclear installation project successfully moves forward and is not backtracked at a later stage.

During the site characterization stage, part of the site evaluation process that follows site selection stage is undertaken. The requirements established in SSR-1 [2] are supported by recommendations provided in associated Safety Guides [3–9].

Requirement 3 of SSR-1 [2] states:

> **"The scope of the site evaluation shall encompass factors relating to the site and factors relating to the interaction between the site and the installation, for all operational states and accident conditions, including accidents that could warrant emergency response actions."**

Requirement 14 of SSR-1 [2] states:

> **"The data necessary to perform an assessment of natural and human induced external hazards and to assess both the impact of the environment on the safety of the nuclear installation and the impact of the nuclear installation on people and the environment shall be collected."**

In addition, SSR-1 [2] states:

> "4.44. Data on natural and human induced external hazards with the potential to affect the safety of the nuclear installation shall be collected throughout the lifetime of the nuclear installation. Data shall be confirmed to be relevant (spatially and temporally) to the site, with preference given to the use of site specific data in site evaluation.
>
> "4.45. The extent, objectives and scope of the data collection process shall be defined on the basis of the safety objective for site evaluation and shall be commensurate with the hazard posed by the nuclear installation to people and the environment.
>
> "4.46. At a minimum, the data collection process shall include the following:
>
> (a) Information on natural and human induced external hazards, including information on sources of hazards, propagation of hazards and the potential effects on the nuclear installation and on people and the environment;

(b) Information describing site conditions and regional environmental conditions;
(c) Information on the proposed engineering and administrative measures for site protection and mitigatory measures;
(d) Information on the potential impact of the nuclear installation on people and the environment for operational states and accident conditions;
(e) Information needed for planning effective emergency response actions on the site and off the site in all environmental conditions and for all states of the nuclear installation;
(f) Information on conditions for access to the site and information for supporting design and development of the site infrastructure.

"4.47. Information and records, if available, of the occurrence and severity of important prehistoric, historical and recent natural phenomena shall be obtained as appropriate for the hazard to be evaluated and shall be analysed for reliability, accuracy, temporal and spatial relevance, and completeness.

"4.48. The data shall be maintained and reviewed periodically, and/or as necessary as part of a review of the site evaluation within the framework of the periodic safety review of the nuclear installation, for example, to address developments in data gathering techniques and in the analysis and use of data and to confirm that the data remain relevant to the site within the context of evolving hazards.

"4.49. The data collected for site investigations shall be of sufficient quality and quantity to support the selected methodology for hazard evaluation.

"4.50. The details of the information collected for each hazard shall be appropriate for the distance between the source of the hazard and the site and the potential impact on the site. The sources of uncertainties relating to data collection shall be documented."

The suitability of the site has to be clearly demonstrated during the early part of the site characterization stage by carrying out all necessary detailed studies so that the huge amount of site related work involving cost and time implications during the site characterization stage can be executed smoothly.

Requirement 4 of SSR-1 [2] states:

**"The suitability of the site shall be assessed at an early stage of the site evaluation and shall be confirmed for the lifetime of the planned nuclear installation."**

In addition, SSR-1 [2] states:

"4.6. In the assessment of the suitability of a site for a nuclear installation, the following aspects shall be addressed at an early stage of the site evaluation:

(a) The effects of natural and human induced external events occurring in the region that might affect the site;
(b) The characteristics of the site and its environment that could influence the transfer of radioactive material released from the nuclear installation to people and to the environment;
(c) The population density, population distribution and other characteristics of the external zone[1], in so far as these could affect the feasibility of planning effective emergency response actions, and the need to evaluate the risk to individuals and to the population.

"4.7. The site shall be deemed unsuitable for a nuclear installation if one or more of the three aspects listed in para. 4.6 indicates that the site is unacceptable and the deficiencies cannot be

[1] The external zone is the area immediately surrounding a proposed site area in which population distribution and density, and land and water uses, are considered with respect to their impact on planning effective emergency response actions.

compensated for by means of a combination of measures for site protection, design features of the nuclear installation and administrative procedures.

“4.8. Site suitability shall be assessed on the basis of relevant current data and methodologies. If relevant, conservative criteria shall be developed in relation to site specific accident scenarios, and the consistency of such criteria with the overall site suitability shall be demonstrated.

“4.9. A decision regarding the suitability of the site shall be based on the characteristics of the nuclear installation, including planned operations at the site, the amount and nature of potential radioactive releases and their impact on people and the environment.”

The Specific Safety Guides [3–9] provide more detailed recommendations on the collection of data for each specific hazard, on methodologies for carrying out the analyses, and establish the suitability of a nuclear installation site with exclusion criteria.

The data that need to be collected during the different stages of site evaluation for a nuclear installation to meet the requirements of SSR-1 [2] are as follows:

- — General characteristics of the site and surrounding region that could affect the safety of the nuclear installation and its radiological impact on people and environment. The sizes of the investigated regions are defined for different natural and human induced external hazards, considering present and foreseeable future characteristics. Paragraph 4.14 of SSR-1 [2] states that "Both the magnitude of the hazard and the distance from the source of the hazard to the site shall be considered in determining the size of the region to be investigated.” Available information collected during the site survey and selection stages will provide a good initial database;
- — General data needed to screen out, identify and establish a list of potential external hazards related to natural phenomena, human induced events and human activities that could affect the nuclear installation. In accordance with Requirement 6 of SSR-1 [2], external hazards not excluded by screening are to be evaluated and used in establishing the site specific design parameters. Furthermore, data are needed (Requirement 10 of SSR-1 [2]) on the potential changes over time that could modify such a list;
- — Data allowing evaluation of the consequences of the listed external natural hazards and human induced events, considering frequency and severity, and potential combinations of external events, in establishing site specific design parameters. Paragraph 4.22 of SSR-1 [2] states that “Appropriate methods, supported by numerical models, when necessary, shall be used to characterize the hazards relevant for site evaluation and design of the nuclear installation.”;
- — Data needed to meet Requirement 11 of SSR-1 [2] on special considerations for the ultimate heat sink for nuclear installations that require an ultimate heat sink;
- — Data allowing evaluation of the consequences of cumulative events;
- — Data necessary to meet Requirement 12 of SSR-1 [2] for evaluating the radiological impact on the public and the environment during normal operation and accidents;
- — Data necessary to meet Requirement 13 of SSR-1 [2] for establishing feasibility of planning effective emergency response actions in case of external events;
- — Any other data necessary to meet Requirement 8 of SSR-1 [2] for adapting the design of the nuclear installation to the site characteristics and establishing measures for site protection.

The necessary databases to comply with above mentioned safety requirements of SSR-1 [2] are described in Annex II, in the context of the safety, as well as with regard to environmental and socioeconomic aspects [3] in the different natural compartments of a site, i.e. the geosphere, hydrosphere, atmosphere, biosphere and anthroposphere. The purpose of gathering data for each of the above site related topics are detailed in Annex II.

The data relevant for the site will include ranges of uncertainty, accounting for spatial variability both at the site and in the surrounding area (see example on vibratory spatial variability in SSG-9 (Rev. 1) [6]). All this information must be provided in the reports required by the regulations during different phases of a nuclear installation project, e.g. site evaluation report, EIA, PSAR and FSAR.

This collection of data on site characteristics increases progressively with the project development in order to meet different requirements of analysis and studies with different levels of precision and details. This approach also must ensure compliance with specific requirements of the regulatory body and other stakeholders through the site survey, site selection, characterization, pre-operational and operation stages and finally, decommissioning of the nuclear installation. Such data collection begins with the already available data and knowledge (e.g. literature, administrative information, public or private companies working on geography, geology, hydrology) during the site survey stage (Section 2.3), then potentially enriched by additional data, limited fieldwork (e.g. mapping and specific site surveys (e.g. geophysical)) during the site selection stage (Section 2.4), substantial enhancement from field work during the site characterization stage (Section 2.5), followed by on-site monitoring both before and during the operational stage of the nuclear installation. Depending on the objectives, data may need to cover the whole site area or even a larger area.

The types and the amount of site data to fulfil the regulatory process increase during the generation of the site evaluation and EIA reports, the application for site approval, the PSAR for a construction licence and finally the FSAR for commissioning of the nuclear installation. The site characteristics are covered in a chapter of the PSAR, with more details included in the FSAR after construction is completed. This chapter provides information on geography, population distribution, geological, hydrogeological, seismological, volcanological, geotechnical, meteorological and hydrological characteristics and human induced events for the site and region surrounding the site. Other chapters of the safety analysis report as well as various sections in the EIA report also provide information on the topography and geography, atmospheric, aquatic and terrestrial environment of the site, demography, land use, health of the population, as well as other industrial and military activities and transport infrastructure in the area. These topics are detailed in Annex II.

For each hazard associated with natural phenomena and/or human induced events mentioned in this section as well as in Annex II, the quantity and types of data needed may appear to be large. However, typically there is a trade-off between the effort and time required to compile a comprehensive, reliable and related database and the level of uncertainty that needs to be considered at various steps of the process [3]. The quantity of data needed and associated level of investigations on the site depend both on the site characteristics (and pre-existing knowledge) and on the type of nuclear installation to be developed (see Section 3.2 below), notably the type and power of the nuclear installation. The use of a graded approach is discussed in Section 4. When the consequences of such hazards, accounting for the remaining uncertainties, are acceptable in terms of effect on health of people and the environment, the continuation of data collection could be limited to the site monitoring.

The siting process for a nuclear installation is significantly shaped by specific considerations for environmental protection of the region and typically involves protecting air, water, cultural resources and wildlife as indicated in Ref. [12]. The complexity and duration needed to obtain the necessary environmental permits can vary significantly among sites, reactor types and individual States. Hence, selecting a site with fewer potential environmental impacts can both accelerate the permitting process and potentially decrease construction and operational costs.

## 3.2. CONSIDERATIONS TO OPTIMIZE THE VOLUME OF DATA NECESSARY FOR SITE EVALUATION

Following subsection provides details on different factors important for optimizing volume of data for site evaluation of a nuclear installation.

### 3.2.1. General

The volume of site data to be acquired for a nuclear installation for site evaluation can be optimized, especially for advanced NPPs. The actual data needed around a nuclear installation site, based on its characteristics depends on (a) from how far an external hazard can affect a site, and (b) how far a nuclear installation can affect the site environment through different mediums during operational states and accident conditions.

In order to reduce or optimize the volume of site data, important factors include pre-selecting regions with suitable site characteristics, the type of the nuclear installation project, the already available data of the site area and the nature and the degree of complexity of the site. These aspects are discussed in the following subsections.

### 3.2.2. Pre-site selection considerations

The main objective of siting for a nuclear installation is to select one of the most suitable sites in a country with site characteristics that result in avoiding or reducing the impact of significant hazards at the selected site. More hazards associated with natural phenomena and human induced events affecting a site means more data must be collected. This will not only result in either shorter or longer data collection efforts but will also affect the site evaluation time. This subsection explains how the site characteristics influence data collection efforts.

Siting issues and efforts can be substantially reduced in terms of data collection and site evaluation if the following are considered during the early stages of site survey:

- Seismic hazard: Seismic hazard evaluation at the nuclear installation site is one of the most time-consuming tasks as data collection and evaluation require extensive effort. Lower seismicity regions with available geological, geophysical, geotechnical and seismological data will require much less time for collection of missing information and/or additional data as compared to high seismic regions with complicated tectonics and poor available data. Moreover, fewer seismic sources around a site means reduced volume of data collection;
- Volcanic hazard: During initial assessment, the volcanic hazard in the site region might be screened out for further evaluation by using appropriate screening distance values as recommended in SSG-21, Volcanic Hazards in Site Evaluation for Nuclear Installations [9]. The project may encounter substantial delays if the potential of future volcanic activity in region surrounding the site cannot be ruled out. This may lead to extensive data collection and a prolonged evaluation of potential volcanic hazards, along with significant cost implications;
- Geotechnical hazard: Rock sites require much less time for data collection, as the quantum of drilling and testing is limited, and generally there are no subsurface hazards. However, softer, sandy soil with a high water table may have liquefaction potential, and clayey soils may have concerns such as consolidation, and they and are difficult to dewater during the construction period. Such soil sites require more dense drilling grids, greater drilling depths and extensive testing;
- Flood hazard: River sites or coastal sites with potential flood hazards will need more time for data collection and evaluation. If no suitable site near water is available, and depending on the cooling water requirements (perhaps dry cooling or non-light water reactor designs are feasible), the nuclear installation need not to be located close to a major water source;
- Meteorological hazard: Data on meteorological events is generally available from regional meteorological stations, but additional data collection and analysis may be extensive if the site is far from existing monitoring stations. Moreover, if a site is subject to extreme, rare events such as tornadoes, further data collection and analysis may be required;
- Emergency planning zones: Since one of the possible advantages of advanced reactor designs is that these could be located close to a population centre, due to much smaller radiological risk, sites with suitable meteorological conditions leading to smaller EPZs are preferred. Sites with high average wind velocities and less stable atmospheric conditions result in smaller EPZs as compared to sites with low average wind velocities and stable conditions. At least two feasible evacuation routes from the EPZ must be ensured during an emergency;
- Hydrogeological aspects: Rock sites with a deep water table and with less data collection needs are preferred to soil sites with a high water table, which need hydrogeological studies and extensive monitoring of groundwater. Furthermore, a high water table may require dewatering during construction, leading to additional costs and longer construction periods;

— Human induced external events: Most human induced hazards within 10 km of a nuclear installation should be considered [4]. As such, the data collection efforts will be reduced if a site with fewer such hazards is selected.

By considering the above site attributes, suitable regions can be identified that will result in selecting a site with minimum data collection efforts and thereby reducing the total site evaluation time and cost.

Applying a graded approach to the siting of nuclear installations other than NPPs can lead to reduced efforts on data collection. A graded approach is generally not possible for high risk installations such as large NPPs. Many advanced NPPs pose less risk to their surroundings, so application of a graded approach is possible.

Advanced NPPs may provide benefits such as reduced initial capital investment, increased scalability, greater flexibility in siting, improved safety and security compared to earlier designs, limited on-site preparation and reduced construction periods. As such, site evaluation for advanced NPPs has also to be optimized in terms of cost and time. This is possible if suitable regions during the pre-site survey stage are identified.

Prior to starting the site survey campaigns, it is important to collect information on different types of advanced NPP technology. There is great variability in their designs and features. In particular, a reduced need for water, the availability of non-water designs, a reduced EPZ and the possibility to reuse brownfield sites (including nuclear sites) need to be considered.

### 3.2.3. The type of the nuclear installation

For site evaluation of nuclear installations other than NPPs, depending on the likelihood that the nuclear installation will give rise to radiological consequences, a graded approach can be applied. This likelihood will depend on the specific features of the nuclear installation, notably the amount, type and status of the radioactive inventory and its distribution, the intrinsic hazards associated with the nuclear physical processes, the configuration of the installation as well as characteristics of the processes or of the engineering features.

SSR-1 [2] states:

> "4.1. The scope of the site evaluation shall cover all external hazards, monitoring activities and site specific parameters relevant for the safety of the nuclear installation. In determining the scope of the site evaluation, a graded approach shall be applied commensurate with the radiation risk posed to people and the environment.
>
> "4.2. The application of the safety requirements for site evaluation for nuclear installations shall be commensurate with the potential hazards associated with the nuclear installation.
>
> "4.3. The level of detail needed in the evaluation of a site for a nuclear installation shall be commensurate with the risk associated with the nuclear installation and the site and will differ depending on the type of nuclear installation.
>
> "4.4. The scope and level of detail of the site evaluation process necessary to support the safety demonstration for the nuclear installation shall be determined in accordance with a graded approach."

In view of above, site investigations for advanced NPPs may be adapted to their associated risks, taking into account these considerations. This practice has been followed by States with low-risk nuclear installations such as research reactors through application of a graded approach. IAEA Specific Safety Guides also provide recommendations on using a graded approach for intermediate hazard category installations by using simplified methods with a restricted database (e.g. SSG-18, Meteorological and Hydrological Hazards in Site Evaluation for Nuclear Installations [5], SSG-9 (Rev. 1) [6]). An example of the application of a graded approach to seismic hazard analysis is also provided in Appendix A.

Other considerations related to the type of nuclear installation are:

(a) The potential consequences both on and off the site in case of an accident are required to be taken into account (see para. 4.5 of SSR-1 [2]). There is a possibility that advanced NPP designs pose a much smaller radiological risk to the people and the environment than existing NPPs. If it can be shown with a high degree of confidence from the beginning of the project development that no off-site release will occur in the most extreme accidental situations (similar to very low power and/or subcritical nuclear facilities), the volume of the data needed for planning and implementing an emergency plan will be minimized. This assumption may be valid for a single SMR unit. However, an installation with multiple SMR units may have a similar EPZ, if the design ensures that failure of each module is independent, i.e. common cause failures due to external hazards or common safety features (such as reactor pool, control room and spent fuel pool) are essentially eliminated;
(b) The collection and processing of population data for an advanced NPP is determined primarily by the size of the EPZ. This zone may be significantly smaller than that for a large NPP. Nevertheless, this depends on the intended number of SMR units and whether sufficient justification is provided to assume single failure module only, similar to the previous case above. Another aspect to consider is the extent of site data needed to identify a cooling source or possibly a water body for effluent discharge. This implies the proximity of a river, lake or ocean, and therefore an increased risk of flooding. In addition, the safety dependency of the installation on a water source may need to confirm the durability and quality of this source for cooling (e.g. temperature, flow rate, absence of debris). For an installation that does not need such a source (e.g. using non-light water reactor (LWR) cooling), a site may be sought that is far enough away from watercourses to reduce the number of site features to be investigated;
(c) One appeal of SMR installations is their ability to be placed as close to a population centre or close to an isolated facility using the energy. Conventional NPPs are often located far from the beneficiaries and supplied power by long distance transmission lines. The proximity of a facility to human activities requires detailed information on the distribution of the population, its lifestyles (e.g. food, leisure, transport) and industrial and military activity around the site. For an installation that does not require such proximity, a greater distance from population centres may be sought in order to limit data collection needs.

### 3.2.4. Issues related to data and information for different types of sites

The level of investigative detail that needs to be carried out by the future operating organization depends on the data and/or information already available at any identified site. Detailed information already exists for sites already hosting nuclear activities but may not be sufficient to meet the requirements of a different type of a nuclear installation and current regulations. For advanced reactors, three types of sites are generally available: [16]: existing nuclear power plant sites, brownfield sites and greenfield sites. For greenfield sites, no known history of industrial development has occurred, whereas brownfield sites typically consist of former commercial and/or industrial lands. Once the site type is known, its site characteristics are assessed and classified as either acceptable, requiring further evaluation, or unacceptable for licensing the advanced NPP technology being considered.

Locations at or adjacent to an operational or former nuclear power plant have the benefit of using previous site characterization studies conducted with the necessary degree of accuracy and precision. Use of such information can greatly assist an advanced NPP site evaluation, and subsequent site assessments can be expedited. Moreover, existing NPP sites possesses the advantages of prior development of related site infrastructure and an existing connection with the electrical grid. Hence, data collection efforts will be minimized with this option.

A brownfield site is an area that has been used for industrial or urban activities in the past or is currently used for such purposes. A brownfield site has both advantages and disadvantages. These sites, such as those previously used for an older thermal power plant, have the advantage of prior development, are more likely to be of a size suitable for advanced NPP development, and they have an existing grid connection adequate for a modular plant, reducing the ancillary costs associated with greenfield site development. A brownfield site may have had past in-depth site characterization, but the data may be limited or of inadequate quality to meet current regulatory requirements.

Similar issues may arise with existing information for a site that has been explored for natural resource use or extraction. If a site has a geological map at a local scale (1/50 000 or more), further mapping may be unnecessary unless certain structures (such as faults) require additional investigation. The investigative drillings carried out and their type (mineral, energy, groundwater or scientific investigations) provide useful information on the subsurface strata and conditions. However, prior investigation could potentially be detrimental, since it could indicate the presence of local natural resources of interest. Such resources, if proven and incompatible with the desired nuclear activity, can lead to the rejection of the site. If these drillings reveal suspected natural resources (potential interest but not exploited to date), it may be necessary to carry out other drillings to confirm either the absence of interest or the compatibility of the nuclear activity with the extraction of such resources.

A possible disadvantage of a brownfield site is potential complications caused by the presence or migration of hazardous substances, pollutants or contaminants from previous activities. The presence of undesirable contaminants at a brownfield site may need to be investigated and resolved. Any nearby industrial facilities may also pose some level of risk [16]. The choice of a brownfield site for a nuclear installation location is usually not straightforward. It is site dependent and can be either be favourable or unfavourable.

Greenfield sites are typically situated outside of urban and industrial areas, often in agricultural or forested land, or other undeveloped areas not known to be polluted. Developing such sites generally involves fewer concerns about prior activities but often lacks site specific data for conducting a thorough initial site evaluation. Assessing whether a greenfield site is suitable for an advanced NPP may present higher levels of uncertainty, particularly due to the potential risks to pristine environments and protected indigenous species [16]. Hence, an extended and more detailed scope of site investigations and characterization studies are needed, leading to increased data collection efforts.

### 3.2.5. The degree of complexity of the site

A site located in a region with high topographic variations is generally a complex site, in particular for significant and varied geotechnical hazards. Steep slopes require an analysis of their stability (e.g. landslides) but also of the flow of runoff water during episodes of heavy rainfall. Topographical variations also cause local meteorological phenomena (e.g. rains, winds) requiring local investigations, as the recordings on existing meteorological stations in the near-regional areas may not be representative.

Several other features of topographically variable areas require more investigation than a flat area. For example:

- The variability of human occupations and activities, to be considered in the impact assessment.
- The possible greater diversity of outcropping geological layers and therefore of soils to be characterized;
- Possibly greater hydrological and hydrogeological complexity that is difficult to model.
- The seismic hazard made more complex by site effects;
- Transportation of massive plant equipment to the site can be problematic as certain parts can be overstressed if transported at higher slope than allowed;
- Dispersion studies are complicated and need large set of data. The selection of an atmospheric dispersion model needs to be aligned with the complexity of the modelling domain, as basic Gaussian models don’t adequately represent dispersion in complex terrain;
- A possible weak transport network, which could hamper emergency evacuation. Sites isolated by mountains or located on a peninsula or small island can also have challenges in developing an emergency evacuation plan.

Site geology greatly influences the volume of data to be acquired. The primary factor is the degree of structural complexity. Sedimentary basins are easier to characterize than regions close to mountains, deformed stratigraphy and igneous intrusions. Ideally, a site in a sedimentary environment with horizontal or gently dipping strata, homogeneous from a sedimentary point of view and not deformed or faulted by more than a few metres, may only require a single drilling phase of investigations to be characterized.

It is also advisable to seek a simple hydrogeological setting, as it may be easier to model and represent the pathways through which radioactive material accidentally released from a nuclear installation could disperse and be transported to areas where they might impact the public. Its complexity is directly linked to the topography and geology of the site. In a simple hydrogeological setting, the hydrological outlets and the transport times to these outlets are generally easily determined from a few piezometric measurements. However, certain features can complicate the hydrogeological scheme:

- Faults and joints that can serve as fast pathways connecting a high level aquifer with deeper aquifers potentially used for drinking;
- Karst hydrogeology that can significantly accelerate transfer times to outlets.

The data needed to assess natural hazards primarily depends on the level of risk in the region of interest. In many regions, volcanic hazards can be easily discounted (and thus the related data collection dispensed with) during the site selection phase. With regard to flooding, the ideal site (i.e. less data needed and lower construction costs) is a dry site capable of withstanding the worst case flooding without the use of external or incorporated barriers (e.g. Ref. [16]).

In view of the above, sites with variable topography are more complex, and more detailed information is necessary for characterization.

## 4. GRADED APPROACHES TO SITE CHARACTERIZATION AND EVALUATION

This section outlies a general approach for applying a graded approach to site characterization nuclear installations. This approach is based on risk-informed and performance-based principles and is intended for advanced NPP designs incorporating new safety features. The main consideration is that performance targets corresponding to large traditional NPPs can be relaxed for reactors that pose an inherently smaller radiological hazard, assuming that the smaller hazard has been appropriately assessed and supported. With regard to this, a detailed graded approach is described, with the main emphasis on the activities that involve the most effort, e.g. the seismic hazard analysis and the geotechnical investigation. A PSA based graded approach is also discussed in this section.

### 4.1. OVERVIEW

#### 4.1.1. Current context

IAEA safety standards dealing with site characterization, external hazard assessment and design against external hazards first provide requirements and recommendations for traditional, land based, NPPs. The standards go on to state that a graded approach can be applied to other nuclear installations. The application of a graded approach is an integral part of Principle 5 of SF-1 [1]; however, practical guidance on how this could be performed is needed.

It is important to note that there is no single method for applying a graded approach; multiple approaches can be employed, each of which may be equally valid depending on specific circumstances. The only condition is that the relevant safety requirements are met. Those requirements can ultimately be expressed as maximum acceptable doses to workers and to the public, even though, for simplification purposes, they are usually introduced in the design and safety assessment framework using surrogate criteria. In this context, the maximum annual probability of exceedance for a design basis earthquake or the maximum acceptable seismically induced core damage frequency can be considered as a part of a set of surrogate requirements[2].

The surrogate requirements have been tuned based on the large body of experience available for traditional NPPs (e.g. LWRs), to ensure they meet the safety requirements for protection of people and the environment from harmful effects of ionizing radiation. However, the surrogate requirements for traditional NPPs may be overly conservative and, in some cases, unsuitable for other, less hazardous nuclear installations. This concept forms the basis for the application of a graded approach.

At present, an increasingly large number of standardized NPP designs are available, which have already been designed for generic site conditions that try to envelop the conditions to be found in most of the potential NPP sites [17]. This means that structures, systems and components (SSCs) have been assigned a safety class and designed accordingly, using nuclear grade standards.

When a site is selected, confirmation is needed that the generic site conditions used for the design (e.g. the design basis earthquake) do envelop the actual site conditions. For this purpose, the selected site needs to be investigated and characterized.

Site investigations and site hazard assessment for a traditional NPP can be time consuming and require significant effort. For example, a site specific seismic hazard analysis for a new NPP following current regulatory requirements in some States may take an average of three years[3].

Since site investigation and hazard assessment requirements for a new NPP site may be considered as surrogates, which have been essentially tuned to generation III NPPs (e.g. LWRs), a question then arises about the possibility of applying a graded approach to advanced NPP designs that incorporate inherent safety features. Those safety features could make the facility intrinsically safer than traditional NPPs and, therefore, there is a possibility that site investigations and hazard assessment requirements for traditional NPPs may be excessively conservative for meeting the same ultimate safety goals for advanced NPP designs.

The collection of environmental data to support site suitability assessments and environmental reviews could also be reduced (see example in Ref. [19]). The smaller the area potentially impacted by a nuclear installation, the lesser the data needed on ecosystem components (e.g. water, aquatic and/or terrestrial species, flora). Reactors with inherent safety features will be likely to have smaller releases and lesser impacts on the environment. Impacts might be limited to the impacts on the installation itself.

### 4.1.2. Bases for applying a graded approach

In general terms, the risk posed by a nuclear installation can be conceptually expressed by the product of the annual frequency of failure to meet the intended safety functions and the consequences of such a failure. High level performance targets are the same for all nuclear installations within the same State. This is a basic safety principle. The performance targets cannot be graded.

Considering just external events, if we assume that failure is the release of radioactive material after fundamental safety functions[4] are lost, the annual frequency of failure against an external hazard depends on:

(a) The site hazard, expressed as an annual frequency of exceeding given severity thresholds of the external events;

---

[2] To illustrate the idea, in some States, it is considered adequate that the design basis earthquake for a nuclear power plant is set at an annual frequency of exceedance of about $10^{-4}$ per year. This is a surrogate criterion for not having unacceptable doses to the public exceeding the annual frequency established in the State. The final goal is not having unacceptable doses, rather than not exceeding the design basis earthquake with a given annual frequency.

[3] This is the average duration of a PSHA study following the SSHAC Level 3 protocol. Reducing the level of the SSHAC protocol to 1, roughly reduces the duration by a factor of three and the cost by a factor of ten [18].

[4] Fundamental safety functions for nuclear power plants are: (a) Control of reactivity; (b) Cooling of radioactive material; and (c) Confinement of radioactive material [20].

(b) The conditional probability of failure for maintaining the fundamental safety functions for an event of a given severity;
(c) The probability of release, for different amounts of radioactivity, once the fundamental safety functions are lost.

If, on the other hand, we characterize the consequences of failure by the doses received by the public and the workers, those consequences depend on:

(d) The amount of radioactivity that is released;
(e) The meteorological/hydrological conditions at the site;
(f) The site configuration/population distribution around the site.

Typically for NPPs, points (a)–(c) above have surrogate requirements to limit the annual frequency of releases to below $10^{-6}$ (see para. 27 of Ref. [21]). For instance, the design basis earthquake is set in an annual frequency range of $10^{-4}$ to $10^{-5}$, and the seismic design provides a probability of failure less than 1% for the design basis earthquake [22]. The threshold on annual frequency of release is in practice a surrogate goal for the ultimate risk goal expressed as frequency of exceedance of allowable doses.

In practice, points (d)–(f) above are checked deterministically (accident analysis) using a design dependent set of source terms and containment leak rates, together with site dependent meteorological and hydrological conditions. These analyses take into account the safety features introduced by the design for mitigation and arrest of design accidents (e.g. safety injection systems, containment spray systems). The nuclear regulations of the State define maximum acceptable doses for workers and the public derived from these postulated source terms.

Looking at the whole set of elements that contribute to risk from external events, in the case of new advanced NPPs, it is possible that a smaller size and the inherent and passive safety features integrated into the design lead to a reduction in points (c) and (d) above, which could allow to a relaxation of points (a) and (b), always keeping the same risk goal as for NPPs. For instance, the plant could perhaps be designed for a maximum seismic event less severe event than the $10^{-4}$ to $10^{-5}$ annual frequency event, using less demanding seismic performance criteria for the design.

In other words, a reduction in the reliability[5] of the nuclear installation (i.e. combination of points (a) and (b) above) might be acceptable, provided that overarching risk goals (e.g. acceptable doses to the public) are respected, as a result of a reduction of the consequences of failure (i.e. points (c) and (d) above). For seismic design, this idea has been in the standards of some States for over a decade, and it is the basis of the so-called performance goal based (risk informed) approach [23–27].

### 4.1.3. Applying a graded approach

Summarizing the ideas presented above, the graded approaches described in the remaining subsections of this section are based on the following general rules:

(a) There may be several acceptable methods of applying a graded approach. This publication suggests some ways regarding site investigation and characterization, but there may be other equally valid ways;
(b) Risk goals cannot be graded. They must remain constant for all nuclear installations within a given State. Risk goals are ultimately expressed as annual frequency of exceeding acceptable doses for workers or the public. For simplification purposes, risk goals are usually introduced in the design and safety assessment framework using surrogate objectives or conditions (e.g. annual frequency of exceedance of the design earthquake). The surrogate objectives have been tuned based on the large body of experience available for traditional NPPs (e.g. LWRs) and may be overly conservative for other types of nuclear installations;

[5] In the present context, reliability may be understood as the probability that the nuclear installation meets its fundamental safety functions for a specified period.

(c) Risk has two components: (i) the frequency of failure at the installation, and (ii) the (conditional) probability that, once failure occurs, workers or the public receive a dose exceeding the acceptability thresholds. Hence, to assess the risk, the site installation interaction needs to be considered;
(d) Point (c.i) above is the reliability of the installation and it depends, among other things, on how the site investigation is performed and on how design basis parameters are derived from it. The application of a graded approach to site investigation affects the level of uncertainty in parameters characterizing the site;
(e) A reduction of the reliability with respect to a non-graded approach may be acceptable, as far as the consequences of the failure, point (c.ii) above, are smaller;
(f) Consequences of failures are dependent on the radiological hazard posed by the nuclear installation, which is based on both on the installation design and the site characteristics (e.g. meteorological conditions and population distribution).

The goal is to apply a graded approach to the implementation of requirements for site investigations and assessment of site specific hazard based on the consequences of potential releases.

The application of a graded approach is intended to reduce the effort in site characterization, which may result in larger uncertainties in site parameters. These larger uncertainties must be considered when comparing the site specific hazards to the design basis events used to develop the reactor standard design. The idea is to explore under what conditions the consequences of releases could be assessed, assuming the advanced safety features of the reactor design. The expectation is that some passive safety features could be credited in the consequence analyses if the design margin of those elements against the applicable external events is large enough.

Since the design basis external event for the passive safety features is dependent on the site hazard assessment, the existence of a large enough design margin will need to be assessed by a safety analysis performed at the end steps of the approach.

### 4.1.4. Steps for a graded approach to site evaluation

A graded approach to site evaluation can be applied through the following steps (see Fig. 4):

(1) Perform a radiological consequence analysis for postulated accidents: Using a design dependent set of source terms, perform site specific accident/dispersion analyses to calculate doses for workers and the public, for average conditions (i.e. no conservative bias). The set of source terms could be the same selected for accident analysis in the design [28], if credited safety features are robust enough to survive the postulated event. This robustness will need to be demonstrated later (see step 6 below). Once the source terms are established using core physics and thermohydraulic transient calculations, typically, computer codes such as RADTRAD [29] and XOQDOQ [30] would be used for assessing the doses.
(2) Categorize the nuclear installation within a radiological hazard category: The hazard category depends on the results of the consequence analysis. For high radiological hazard facilities, methodologies for hazard assessment applicable to NPPs are to be used.
(3) Determine if a graded approach is possible: Application of a graded approach depends on the radiological hazard category. For hazard assessment of high hazard category nuclear installations, methodologies for hazard assessment applicable to NPPs are to be used. Application of safety requirements must be similar among NPPs. However, application of a graded approach to safety requirements is possible for nuclear installations categorized as intermediate and low hazard.
(4) Applying a graded approach to site investigation and characterization: This depends on the radiological hazard category. Giving quantitative guidance (e.g. number of samples, number of boreholes, level of detail in hazard assessment process) requires an assumption about what reliability (or frequency of failure) would be acceptable for each radiological hazard category.
(5) Applying a graded approach to site evaluation: Minimum criteria for site evaluation, based on site characterization, need to be set. There is a minimum level of characterization required to assess site acceptability in compliance with the requirements established in SSR-1 [2]. It is not acceptable to go

below this threshold, even though more room could be allowed for judgement based on a minimum set of data.

(6) Derivation of design bases from the graded approach to site characterization: At a single site, design basis external events could be less demanding (i.e. more frequent) for intermediate and low radiological hazard installations. However, less site investigation will lead to more uncertainties, and this needs to be considered when establishing the design bases.

(7) Verification of assumptions made in the consequence analyses (step 1): Robustness of safety features credited in the consequence analyses, if applicable, will translate into a set of requirements for design margin above the design basis events. A safety assessment will need to be performed to demonstrate that the design margin beyond the design basis event is large enough. The minimum acceptable margin to show robustness will need to consider the larger uncertainties introduced by using hazard assessment methodologies simpler than those used for traditional NPPs. If the hazard assessments are being performed for SMRs with multiple units, then it is very important to preclude the failure of more than one module (due to common cause) if the emergency planning zone has been configured for the source term of a single module.

The following sections provide a detailed discussion of these steps.

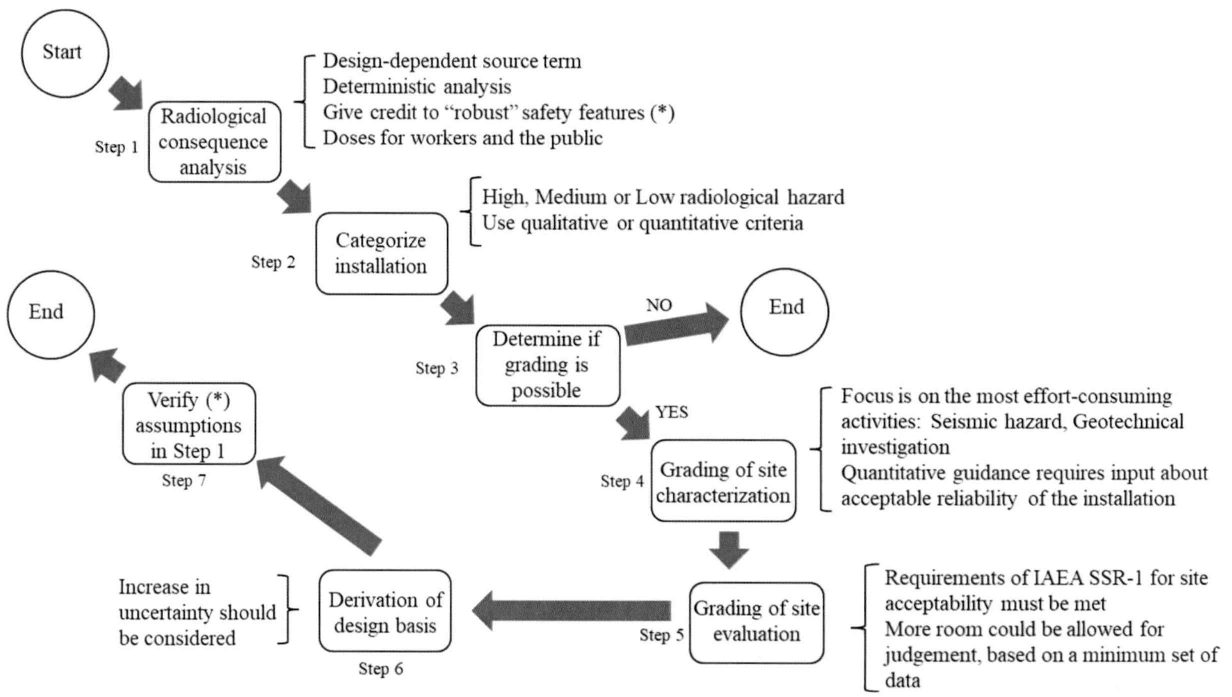


*FIG. 4. General workflow of a graded approach for site characterization.*

## 4.2. CONSEQUENCE ANALYSIS

Application of a graded approach to site characterization has to be informed by a site specific consequence analysis. This analysis, even if simplified, needs to categorize the installation according to its radiological hazard.

In the context of this publication, consequence analysis is the deterministic accident analysis performed to assess radiological consequences in a traditional NPP (see section 15 of IAEA Safety Standards Series No. SSG-61, Format and Content of the Safety Analysis Report for Nuclear Power Plants [31]) but performed for average conditions. Consistent with a risk informed process, the consequence analyses used to categorize an installation in terms of radiological hazard endeavour to use average values for the parameters defining the radioactive release, the dispersion in the environment and the assessment of doses.

Analyses performed in the context of safety assessment of traditional NPPs take into account the safety features introduced by the design for mitigation and arrest of design accidents (e.g. containment spray systems, safety injection systems). A number of design dependent source terms, associated with the performance of engineered safeguards systems, are defined [32], and consequences (doses) for workers and the public are determined based on plant configuration and site specific dispersion analyses.

In contrast, when performing a consequence analysis for the purposes of categorization in terms of radiological hazard, engineered mitigating features are assumed not to function unless the robustness of each mitigating feature can be clearly demonstrated for the postulated event [23, 33, 34].

The simplest consequence analysis involves evaluating the impact of an unmitigated release of the entire radiological inventory from the nuclear installation. This conservative approach offers a preliminary estimate of the nuclear installation's hazard category. If this analysis shows that radiological consequences are minimal for workers, public and the environment, the installation can be categorized to the lowest radiological hazard category. In this case, site investigation and its characterization can be conducted similarly to those for a conventional industrial facility. This approach may be most useful for nuclear installations with a small radioactive inventory (e.g. research reactors). For advanced NPPs, following this approach might lead to the conclusion that the advanced NPP is classified as a high radiological hazard category installation. In view of this, an important input for consequence analyses for radiological hazard categorization of advanced NPPs is identification of safety features that could be credited in order to reduce radioactive releases.

Categorizing the radiological hazard of a nuclear installation based on a consequence analysis is considered acceptable if it uses a design dependent set of source terms and takes into account certain engineered mitigating features. This approach would be deemed acceptable if the source terms comprehensively cover all potential accident scenarios, and the robustness of the mitigating features is clearly demonstrated for design basis events.

In advanced NPPs, passive safety features are expected to have large design margins against external events due to standard design practices. The design robustness of these features can be demonstrated by providing evidence of significant design margin above that required for the design basis event.

## 4.3. RADIOLOGICAL HAZARD CATEGORIZATION

Based on a consequence analyses, four radiological hazard categories can be defined, as shown in Table 1. The radiological hazard categories range from high which is typical for traditional NPPs, to conventional, which pertains to conventional industrial facilities that may pose a negligible or no radiological hazard.

TABLE 1. RADIOLOGICAL HAZARD CATEGORIES BASED ON THE FAILURE CONSEQUENCES OF NUCLEAR INSTALLATIONS
*(modified from Table 2 of Ref. [13])*

| Hazard category | Consequences on site | Consequences off site | Remarks |
|---|---|---|---|
| High | Radiological or other exposures that might cause loss of life of workers in the facility | Potential for significant off-site radiological consequences | Hazard category of traditional NPPs. Site investigation with the same scope as for NPPs. |
| Intermediate | Potential for significant on-site consequences<br>Unmitigated radiological release would necessitate on-site evacuation | Small potential for off-site radiological consequences | |
| Low | Potential for only localized radiological consequences (within 30–100 m of the point of release) | No off-site radiological consequences | |
| Conventional | No radiological consequences | No radiological consequences | Site investigation with the same scope as for conventional industrial facilities. |

**Note**: Quantitative definition of significant or small consequences depends on the regulatory framework of each State.

The radiological consequences due to potential failures depend not only on the nature and design of the installation but also on the characteristics of the site [25]. Therefore, evaluating the radiological hazard of a nuclear installation needs to consider the interaction between the site and the installation. In this regard, following factors need to be considered [13]:

— The amount, type and status of the radioactive inventory at the site (e.g. whether solid, liquid and/or gaseous, and whether the radioactive material is being processed or only stored);
— The intrinsic hazard associated with the physical processes (e.g. nuclear chain reactions) and chemical processes (e.g. for fuel processing purposes) that take place at the installation;
— The thermal power of the nuclear installation, if applicable;
— The configuration of the installation for different kinds of activity;
— The distribution of radioactive sources in the installation;
— The need for active safety systems and/or operator actions for the prevention of accidents and for mitigation of their consequences, and the characteristics of engineered safety features for the prevention of accidents and for mitigation of their consequences (e.g. the containment and containment systems);
— The characteristics of the structures of the nuclear installations, and the means of confinement of radioactive material;
— The characteristics of the processes or of the engineering features that might show a cliff edge effect in the event of an accident;
— The characteristics of the site that are relevant to the consequences of the dispersion of radioactive material to the atmosphere and the hydrosphere (e.g. topography, dominant winds, water masses, demography of the region);
— The potential for on-site and off-site contamination.

## 4.4. APPLICATION OF A GRADED APPROACH TO SITE INVESTIGATION AND CHARACTERIZATION

### 4.4.1. Potential for applying a graded approach

For nuclear installations categorized under the high hazard radiological category (as shown in Table 1), site investigations and characterization need to consider the same scope as conducted for NPPs.

For nuclear installations categorized in the conventional radiological hazard category (as shown in Table 1), the site investigations and characterization scope can be the same as conducted for conventional industrial facilities. The scope of site investigation and characterization does not need to be increased.

A graded approach should be applied to the site investigation and characterization of nuclear installations categorized under intermediate or low radiological hazard categories (as shown in Table 1). This approach involves a reduced level of detail and less rigour compared to that needed for nuclear installations in the high radiological hazard category. Typically, for an intermediate hazard category nuclear installation, it is appropriate to consider a reduction in the scope relative to that used for high hazard category installations. Conversely, for a low hazard category nuclear installation, an enhancement of the scope compared to that used for conventional hazard category installations can be considered [6].

It is advisable to keep the annual probability of exceedance of hazards similar to those with high hazard category and apply a graded approach to the time, effort and resources to be considered for site investigations and analytical work.

### 4.4.2. Applying a graded approach to nuclear installations in intermediate and low hazard radiological categories

While a graded approach can be applied to any site investigation and characterization activities, the primary focus is typically on reducing effort in the most time-consuming tasks. At most sites those activities will be the seismic hazard analysis and the geotechnical investigation and characterization (Table 2). This publication focuses on these two aspects of site characterization. There are other aspects, perhaps not as generally applicable, which could need a significant level of time and effort. This is the case, for instance, for coastal or river flooding.

Table 2 provides approximate time estimates for site characterization of new sites. However, when a new reactor is planned for an existing nuclear site, the process may be expedited. This is because the site investigation would primarily focus on updating existing data to reflect current site conditions and meet the latest licensing requirements, rather than starting from the beginning.

TABLE 2. ESTIMATES OF TIME NEEDED FOR SITE CHARACTERIZATION STUDIES OF A NEW SITE FOR A TYPICAL NPP

| Sr. No. | Name of study | Category | Typical duration for traditional NPPs (years) |
|---|---|---|---|
| 1. | Topography and mapping | Compilation of data | 0.5–1 |
| 2. | Geology and geotechnical conditions | Compilation of data and design parameters | 2–3 |
| 3. | Seismic hazards analysis | Compilation of data and design parameters | 3–4 |
| 4. | Development of seismic monitoring network | Descriptive | 0.5–1 |
| 5. | Volcano hazard analysis | Compilation of data and design parameters | 2–3 |
| 6. | Oceanography and coastal flooding | Compilation of data and design parameters | ~ 2 |
| 7. | Bathymetry and seabed properties | Compilation of data | ~ 1 |
| 8. | Hydrogeological and hydrogeochemical | Compilation of data and design parameters | ~ 2 |
| 9. | Meteorological | Compilation of data and design parameters | 1–2 |
| 10. | Human induced external events | Compilation of data and design parameters | ~ 1 |
| 11. | Demography | Compilation of data | 0.5 |
| 12. | Land, water, marine uses | Compilation of data | 0.5 |
| 13. | Water, electricity, logistics | Assessment based on compiled data | 0.5 |
| 14. | Archaeological and historical monuments | Compilation of data and assessment | 0.5 |
| 15. | Ecology | Compilation of data and assessment | 1.5 |
| 16. | Socioeconomic and social culture | Compilation of data | 0.5 |
| 17. | Dispersion and dose assessment | Assessment based on compiled data | 1–2 |
| 18. | Emergency preparedness | Assessment based on compiled data | ~ 1 |
| 19. | Radiological environmental monitoring | Descriptive | 0.5 |
| 20. | Air quality and noise levels | Compilation of data | 0.5 |
| 21. | Feasibility of security planning | Assessment based on compiled data | |
| 22. | Climate change | Assessment based on compiled data | ~ 1 |
| 23. | Cooling water | Assessment based on compiled data | ~ 1 |
| 24. | Site information management system | Descriptive | 0.5 |
| 25. | Power network integration study | Assessment based on compiled data | 0.5 |

#### *4.4.2.1. Seismic hazard analysis*

This publication considers the application of a graded approach to vibratory ground motion seismic hazard analysis. Fault displacement hazard is outside the scope, as it may involve exclusionary issues and a graded approach cannot be applied. Broadly speaking, a vibratory ground motion seismic hazard analysis for a particular site has three stages:

(a) Compile seismotectonic information relevant for the site. Information comes from existing sources (e.g. technical literature, previous studies) as well as from specific data acquisition campaigns.

(b) Evaluate seismotectonic information and develop seismotectonic models which integrate the information [6]. The models include the characterization of both seismic sources and ground motion (attenuation + site effects).
(c) Process the seismotectonic models to assess the hazard. This can be done deterministically or probabilistically [6]. In both cases, uncertainties need to be considered and accounted for.

The main effort belongs in the first two stages as well as the detail in which the aleatory and epistemic uncertainties are represented. For intermediate and low radiological hazard nuclear installations, a graded approach can be applied to the effort in these two stages. References [33] and [34] provide some general guidance, as described below.

*Level of investigation*

Investigation begins with a review of available information for the site region and a field reconnaissance of the site area. The goal is to develop an understanding of current knowledge about the region and the site conditions. This understanding can afterwards be used to devise an efficient programme of additional data collection. In all cases, compilation and review of information are needed and these include:

- Earthquake catalogues;
- Topographic, geological and geophysical information (maps with cross-sections);
- Aerial photographs and other remote sensing images;
- Digital elevation models, such as those derived from LiDAR data;
- Geological, seismological, geophysical and geotechnical reports or literature covering the study region and site area.

Field reconnaissance mainly in the site vicinity is typically performed after the existing information is reviewed. The main purpose is to identify fault capability issues [6].

The assessment of seismic hazard involves the identification and characterization of seismic sources affecting the site. Seismic sources are either seismogenic structures or zones of diffuse seismicity. Such zones are a discrete volume of the Earth's crust within which earthquakes may occur with uniform probability, but not on currently identified structures.

In moderate and low seismicity areas[6], the investigation needs to focus on characterizing background earthquakes[7] and seismic sources that could contribute more than 5% to the overall hazard. Depending on the seismotectonic setting, this condition could reduce the radius of the investigation significantly under the common value of 300 km used as a reference for traditional NPPs.

For high seismicity areas, in addition, investigation will normally need to characterize in detail all Quaternary faults and volumetric source zones within a 25–40 km radius from the site. In all cases, the site vicinity will need to be investigated for the potential presence of capable faults, since the assessment of capable faults is needed for the safety evaluation of the site [2].

In general, for intermediate and low radiological hazard nuclear installations, the available information will be used to identify and characterize the seismic sources, with no additional data collection activities, unless significant gaps are identified during the review. The compilation of existing information and collection of new data compose the evaluation phase of a seismic hazard analysis. The best available information should be used during this evaluation.

---

[6] Level of seismicity can be assessed from national seismic hazard maps. Generally, an area can be considered of low to moderate seismicity if the 475-year return period hazard map does not predict a 5% damped elastic response spectrum with a peak larger than 0.40 g.

[7] Background earthquakes are those not associated with identified seismogenic structures, but rather a zone of diffuse seismicity.

*Integration in a seismotectonic model*

Following the evaluation phase is the integration phase of the analysis. When performing a seismic hazard analysis, there are uncertainties related to the understanding of earthquake mechanisms and propagation of seismic waves. Current practice distinguishes between two types of uncertainty: aleatory and epistemic. Aleatory uncertainty results from natural variability and randomness in physical processes. Collection of additional data does not help to reduce this type of uncertainty. Epistemic uncertainty arises from incomplete knowledge and can, in theory, be reduced by collecting additional data.

The two types of uncertainty are managed differently in a seismic hazard analysis. Aleatory uncertainty is normally assessed by closed-form mathematical expressions. Typically, integrals are computed over aleatory uncertainties to determine the likelihood that a specific ground motion parameter will be exceeded, based on a particular assumption about the state of nature. In a PSHA, this approach produces a single hazard curve.

Epistemic uncertainty related to this hazard curve is addressed by incorporating various models, assumptions, hypotheses or parameters through logic trees. These multiple interpretations are propagated through the analysis, leading to a range of hazard curves and their associated weights for performing weighted averages. End results are presented as statistical parameter curves of exceedance probability for each amplitude of ground motion, including the mean, median and various fractiles. The central tendency is shown by the mean and median hazard curves. The separation among fractile curves reflects the epistemic uncertainty in seismic source characteristics and prediction of ground motion.

The variation in effort during the integration phase of seismic hazard analysis can be correlated with the evaluation of epistemic uncertainties. The smaller the effort, the greater is the probability of underestimating the epistemic uncertainty (i.e. dispersion around the central or median estimates of the hazard).

For intermediate and low radiological hazard nuclear installations, technical integration could be performed by an individual or small team, with some interaction with peer reviewers. There is no need to bring experts together in workshops to debate and interact on technical issues. If a Senior Seismic Hazard Analysis Committee (SSHAC) approach is being used for the analysis of seismic hazard, then the appropriate level could be selected as either SSHAC Level I or II. SSHAC levels range from I to IV, with higher levels representing progressively more data, analysis and peer review involvement in the analysis, and a resulting higher level of regulatory assurance in the final results. Higher SSHAC levels are appropriate for higher hazard facilities and/or sites with complex or controversial technical issues. Regardless of whether the SSHAC process is applied, or at what level, the entire seismic hazard analysis process needs peer review.

*Resulting uncertainties*

The reduced effort for intermediate and low radiological hazard nuclear installations will likely provide a good representation of median hazard but a poorer representation of uncertainties [35]. This needs to be considered when selecting the design basis earthquake (see Section 4.6.1 and Appendix I).

The trade-off is to be able to finish a PSHA in a timely manner (e.g. ~1 year) and in return provide a robust design for an advanced NPP, taking into account the additional uncertainties in the estimation of both the design basis as well as beyond design basis earthquake.

#### *4.4.2.2. Geotechnical investigation and characterization*

Compared to the scope required for a high hazard category nuclear installation, the application of a graded approach to geotechnical site investigation and characterization is based on the complexity of foundation and subsurface conditions. The potential for applying a graded approach has to be determined by qualified geotechnical engineers.

The following items need to be addressed, at a minimum, when applying a graded approach to geotechnical site investigation and characterization:

— Subsurface materials structure, including a description of the stratigraphic arrangement of subsurface materials, nature and dimensions in plan and depth of the different formations;
— Subsurface materials geotechnical properties including physical, strength (static and dynamic) and chemical properties, for assessing bearing capacity, settlement and stability, and to determine geotechnical related design basis parameters;
— Investigation and characterization of complex subsurface conditions (e.g. underground cavities, expansive soils and rocks);
— Investigation of site hydrogeological conditions (e.g. description of aquifers, groundwater regime, groundwater levels, variations in levels, chemical properties of groundwater).

A graded approach may affect the level of detail (e.g. number of boreholes and quantity and types of laboratory/field tests) for investigating and characterizing these items. However, these items need to be included in scope of the geotechnical site investigation. In addition, variability and uncertainty in characterizing subsurface materials need to be addressed.

A programme for geotechnical site investigation of a nuclear installation based on site specific conditions often cannot be completed in a single phase. Typically, the programme is implemented in multiple phases, with increasing level of detail based on results from each preceding phase. An appropriate graded approach could be used to eliminate or reduce the effort in final phases, where extensive geotechnical work is involved for complete site characterization and final assessment.

Although general rules may prove to be invalid for a particular site, it might be expected that: the number, spacing and depth of site penetrations (boreholes, cone penetrometers); the number and type of soil samples obtained; and the associated testing programme for an intermediate hazard category installation is approximately half the effort for a high hazard category installation (see Appendix II). Annex VI presents an example of such an approach for site investigations for an SMR project.

## 4.5. APPLICATION OF A GRADED APPROACH TO SITE EVALUATION

### 4.5.1. Applicability for the graded approach

For high radiological hazard category nuclear installations (as shown in Table 1), a graded approach cannot be applied to the site evaluation. The same scope as for traditional NPPs needs to be implemented.

For conventional radiological hazard category nuclear installations (as shown in Table 1), site evaluation does not need to follow nuclear safety standards, as the installation pose no significant radiological risk to the public and environment.

For intermediate or low radiological hazard categories nuclear installations (as shown in Table 1), the application of a graded approach to site evaluation should be considered to optimize the level of effort compared to high radiological hazard category nuclear installations.

### 4.5.2. Applying a graded approach to intermediate and low radiological hazard category nuclear installations

#### *4.5.2.1. Seismic hazards*

Site safety evaluation for seismic hazards mainly refers to the fault displacement hazard, especially in relation to the presence of capable faults in the site area or its vicinity [2].

Section 9.13 of SSG-9 (Rev. 1) [6], implies that applying a graded approach to the fault displacement hazard analysis is not appropriate, irrespective of the category of radiological hazard of the installation. As recommended in SSG-9 (Rev. 1) [6], the site investigation needs to provide sufficient data to either rule out surface faulting at or near the site or, if faults are present, it needs to evaluate their capability, in order to allow a confident decision regarding the hazards posed to the nuclear installation and the suitability of the site [3].

#### *4.5.2.2. Geotechnical investigation and characterization*

A graded approach to geotechnical characterization must provide sufficient details for performing site evaluation of geotechnical hazards, such as the potential for slope instability, subsoil liquefaction, subsidence, collapse, or uplift of the site surface. The methods for evaluating geotechnical and geological hazards may follow a graded approach such that, based on site conditions, expert judgement and/or simplified bounding analyses might be sufficient for screening out those hazards.

If the site evaluation does not allow for one or more geotechnical and geological hazards to be screened out, a further detailed investigation and characterization will be necessary for refining the assessment. After refining and conducting further evaluation, the site may be deemed suitable according to site specific established criteria. Design bases then need to be set for ensuring the nuclear installation's safety through proper design, construction, and operational practices.

## 4.6. DERIVATION OF DESIGN BASES

### 4.6.1. Seismic design bases

For vibratory ground motion, the design basis ground motion needs to be selected with consideration of a potentially less rigorous representation of uncertainties in the graded hazard assessment.

A possible approach is to derive a set of corrected mean hazard curves by multiplying the median hazard curves obtained using the graded hazard analysis by suitable factors (Appendix I). Subsequently, the mean hazard curves can be used to derive the design basis earthquake ground motion, following the recommendations provided in SSG-67 [13]. It is important to highlight that the recommendations provided in SSG-67 [13] include not only a reduction in the design basis earthquake ground motion return period for installations in the intermediate or low radiological hazard categories compared to those in the high radiological hazard category, but also advocate for the use of conventional design codes and standards instead of nuclear grade standards.

### 4.6.2. Geotechnical design bases

The outcome of applying a graded approach to geotechnical site characterization is an increased level of uncertainty in the evaluation of geotechnical parameters, which needs to be considered in the design basis for buildings and other civil structures. This larger uncertainty needs to be considered when establishing the site related geotechnical design basis parameters. It will normally result in a reduction of strength parameters with respect to the ones that would have been derived without using a graded approach.

Sometimes, applying a graded approach involves reducing the number and depth of boreholes and limiting other activities such as sampling, field, and laboratory testing. This can be considered a sampling problem, for which confidence intervals for the true mean value can be defined (Appendix II). For instance, the lower end of the 95% confidence interval can be used to determine the design basis parameter. The smaller the number of elements in the sample, the larger the confidence interval will be and, therefore, the smaller will be the lower end of it. In addition, estimates for the standard deviation of the mean value are larger as well.

The application of a graded approach to site characterization may lead to a less detailed understanding of the subsurface materials structure (e.g. variation in different soil profiles at the site), or other characteristics (e.g. physical, geochemical and strength properties of subsurface materials) [36]. Margins in the design basis need to account for these uncertainties by determining reasonable ranges of variation to consider in the design. Alternatively, a conservative approach could involve simply selecting the most unfavourable conditions.

## 4.7. VERIFICATION OF ASSUMPTIONS MADE IN THE CONSEQUENCE ANALYSIS

### 4.7.1. Seismic analysis

In accordance with Ref. [23], for an SSC's mitigating effects to be considered in the accident consequence analysis on which radiological hazard categorization of the installation is based, the SSC has to be identified as robust and be given special attention throughout its entire lifetime (design, procurement, construction, operation and maintenance). For civil structures large design margins have to be provided, typically for a limit state with limited permanent distortion or with essentially elastic behaviour. Limit states and their relation to SSC performance criteria are described in Ref. [23].

The amount of design margin needs to be commensurate with the application of a graded approach to defining the design basis event, with respect to the requirements for a traditional NPP. In addition, the larger level of uncertainty introduced by simplified methodologies for hazard assessment needs to be taken into account. For example, if a nuclear installation is classified as having an intermediate radiological hazard based on the seismic robustness of the containment structure, typically this would imply a design basis earthquake with a return period smaller than the one that would have been used for a traditional NPP (see table 3 in SSG-67 [13]). It would also imply that the site specific hazard would be assessed using a less time-consuming methodology, with the scope for site investigation reduced with respect to the one that would have been used for a traditional NPP. This reduced scope may result in underestimation of uncertainties in the hazard. The design margin to be demonstrated for the containment needs to cover both aspects: the reduction in the return period and the potentially larger uncertainties. It is not possible to both reduce the return period and also simplify the PSHA, and therefore an optimized trade-off needs to be sought.

The first aspect can be addressed using the slope of the mean hazard curves from the annual frequency of exceedance of the design basis earthquake to the design basis earthquake that would have been used for a traditional NPP (see table 3 in SSG-67 [13]). The second aspect can be addressed by considering a conservative increase of the dispersion of the percentiles of hazard (epistemic uncertainty), as mentioned in Section 4.6.1 and described in Appendix I.

Typically, in some States, for high radiological hazard installations, a seismic design margin of between 1.50 and 1.67 times the design basis earthquake must be demonstrated. Therefore, when the seismic robustness of a containment structure is credited for classifying a nuclear installation as low or intermediate hazard, larger margins will need to be shown. For instance, an acceptable margin over a design event at the $4\times10^{-4}$ per year hazard level could be obtained by setting the condition that a margin 1.67 times the design event for the high hazard category ($10^{-4}$ per year) would need to be attained. If the approximate values given in Table 3 are used, then the margin could be computed per equation (2) (e.g. as shown in Ref. [37]):

$$1.67\times\{(10^{-4}/4\times10^{-4})\}^{-(1/KH)}=(\text{low seismicity region})=1.67\times\{(10^{-4}/4\times10^{-4})\}^{-(1/2.75)}=1.67\times1.66=2.77 \quad (2)$$

That is, sufficient margin to consider the containment structure to be robust would be 2.77 times the design event at the $4\times10^{-4}$ per year hazard level. This can be considered an indirect way of checking that the structure would have adequate margin, had the high radiological hazard category event been used for the design.

TABLE 3. TYPICAL GROUND MOTION RATIOS CORRESPONDING TO A TEN-FOLD INCREASE IN THE ANNUAL EXCEEDANCE PROBABILITY [38]

| Seismicity | Annual frequency range ($yr^{-1}$) | Typical ground motion ratio ($A_g$) | Typical slope parameter of mean hazard curve $K_H$ |
|---|---|---|---|
| High (e.g. California) | $10^{-3}$ to $10^{-4}$ | 2.00 | 3.32 |
| | $10^{-4}$ to $10^{-5}$ | 1.67 | 4.49 |
| Low (e.g. eastern United States of America) | $10^{-3}$ to $10^{-4}$ | 2.31 | 2.75 |
| | $10^{-4}$ to $10^{-5}$ | 2.13 | 3.05 |

This example is for illustrative purposes only. The example provides, nevertheless, an order of magnitude of the conditions to be expected. Margins approximately three times the site specific design event are considered to be achievable for structural passive SSCs, especially if the standard plant design was carried out for a design earthquake significantly larger than the site specific design event.

For active SSCs (e.g. safety injection or containment spray systems) large margins may require, in practice, their seismic design withstand an event larger than the site specific event. The consequence is that accident mitigation measures relying on active SSCs may be difficult to be considered robust in the consequence analysis, if the corresponding safety features have been designed for a smaller radiological hazard category.

When a seismic PSA model is available, this tool can be used to directly check the compliance with the performance target established for the nuclear installation, using the corrected hazard curves developed as mentioned in Section 4.6.1. In this case, there would be no need to check the seismic margin of SSCs credited in the consequence analysis, since the capacity of those SSCs would be considered in the seismic PSA via their seismic fragilities.

### 4.7.2. Geotechnical analysis

Derivation of geotechnical design bases, as described in Section 4.4, does not lead to a reduction of the design basis loads. Hence, there is no need for a verification of the assumptions about robustness made in the consequence analysis used to categorize the installation as low or intermediate radiological hazard. This verification must be done whenever a graded approach has resulted in a reduction in the design actions with respect to the actions considered in the high radiological hazard category (e.g. traditional NPPs).

## 4.8. REGULATORY FRAMEWORK FOR REVIEW AND ASSESSMENT

The regulatory effort allocated for the review and assessment of site characterization needs to be consistent with the selected graded approach. The effort needs to be commensurate with the radiological hazard of the nuclear installation [19].

Reference [19] suggests that allocation of resources for the review and assessment be made dependent on the radiological hazard of the nuclear installation. Such an allocation could be made in three steps:

(a) Determine the scope and depth of the review and assessment based on applicable requirements and develop a review plan;
(b) Identify the factors applicable to the decision on resource allocation, including the regulatory body resources, the document submission requirements, the level of detail of information in the submissions, reviews and assessments performed by other regulatory bodies, experience and knowledge, urgency, alternative approaches to those specified in the requirements, novel methods and the margins to safety limits shown by the analyses;

(c) Integrate the identified applicable factors into the decision making about the resources, considering the radiation risks posed by the installation.

If the review and assessment shows that there are non-compliances with requirements, the acceptability of these non-compliances needs to be commensurate with the associated radiological hazard. A determination of the safety significance of the non-compliances will need to use a structured, documented method.

## 4.9. USE OF AVAILABLE PROBABILISTIC SAFETY ASSESSMENT MODELS: SCREENING OF EXTERNAL HAZARDS

### 4.9.1. General considerations

In Sections 4.1–4.8, a general graded approach to siting activities has been presented, which could be applied with no explicit use of PSA for external events. If external event PSA models are available for the standard plant design, they can be used in the assessment of risk, to check compliance with the performance targets associated to each radiological hazard category.

There are two steps in which PSA techniques can play a significant role:

(a) Identification of external hazards with a small contribution to risk, in order to reduce the effort in the assessment of those hazards;
(b) Demonstration that SSCs have enough robustness to meet the installation-level performance targets. This can be done because, in a PSA, the installation-level fragility is calculated from the fragility of the individual SSCs. Therefore, the robustness of each individual SSC is determined and can be used to show that adequate robustness is achieved.

The first point is the main focus of this section. The second point has already been mentioned in Section 4.7.1.

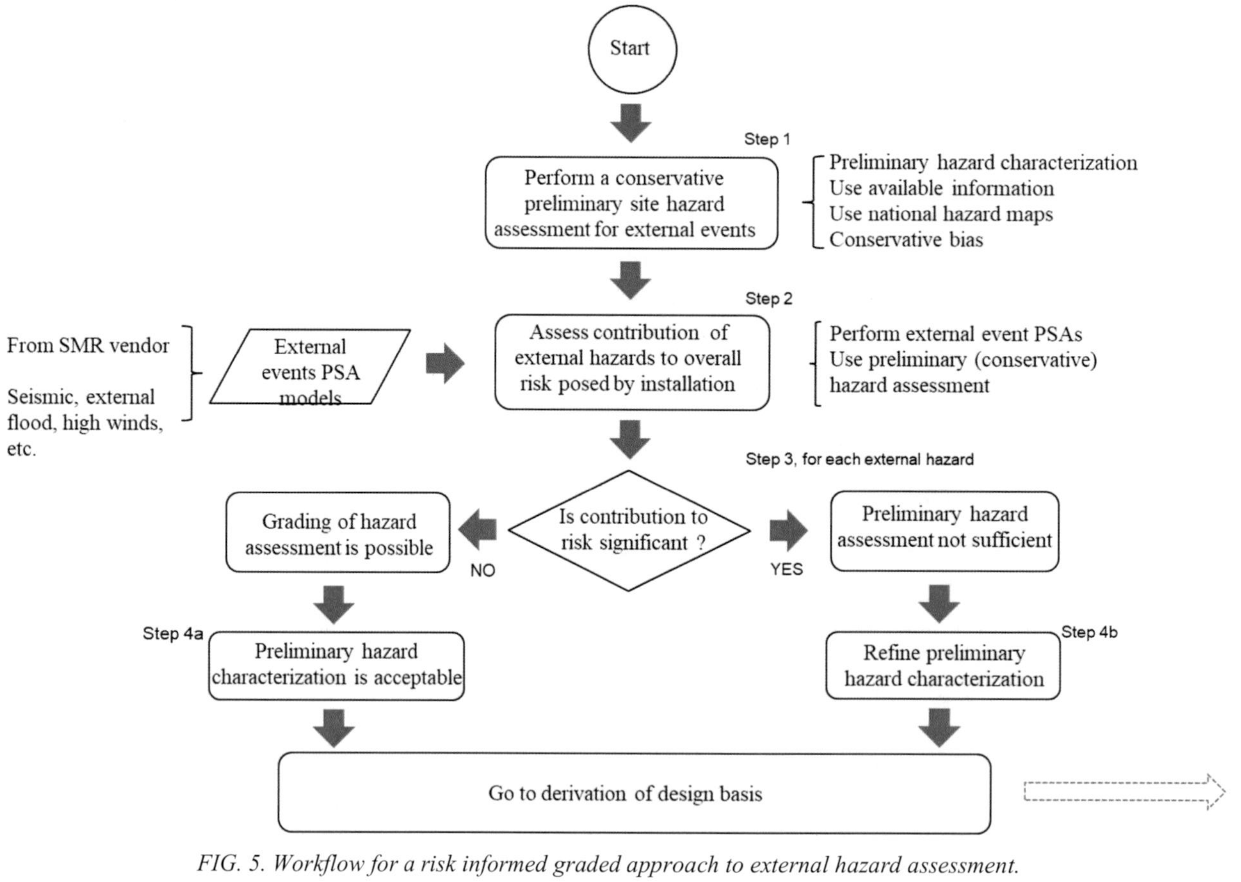


*FIG. 5. Workflow for a risk informed graded approach to external hazard assessment.*

### 4.9.2. Screening of external hazards

Figure 5 schematically shows the process that may be followed to identify the risk relevant hazards. In order to optimize the hazard assessment effort, the effort needs to focus on the risk relevant hazards. Assessment of other hazards may be graded and kept at a preliminary characterization level.

The steps in Fig. 5 are described below: more details can be found in Ref. [39].

(1) Identification of potential hazards and hazard groups at the selected site: From a universe of potential external hazards, possible hazards at the site are identified. See section 2 of Ref. [40].
(2) First screen of potential hazards. This screen is also applied to traditional NPPs (see section 3.2 of SSG-61 [31]). The screen is applied in steps 3 and 4 below.
(3) Based on qualitative criteria, exclude hazards that clearly pose no safety consequences or whose impacts are enveloped by other hazards.
(4) Bounding analyses: Employ simple conservative estimates to demonstrate that the worst case scenario of a specific hazard or a combination of hazards will not lead to safety consequences. It needs preliminary site characterization data, typically gathered during the site selection process.
(5) For the hazards and hazard combinations that have been screened in, develop a preliminary estimate of their annual frequencies of occurrence/exceedance, along with relevant parameters (e.g. loading, magnitude).
(6) For each screened-in hazard, use the external event PSA model to derive applicable risk metrics (core damage frequency, large early release frequency): Quantification of the PSA models requires both an estimate of the hazard (hazard curves) and an estimate of SSC capacities (fragilities) against the hazard. Hence, calculation of SSC fragilities/capacities for each of the hazards is required, even though it may be based on (conservative) estimates. The results of the PSA provide a conservative assessment of the SSCs affected by each hazard, along with the associated failure modes, including those initiating by hazard-induced events.
(7) For each screened in hazard, assess PSA derived risk metrics (core damage frequency, large early release frequency) to determine if the hazard could be a significant risk: Thresholds of risk significance are a subject of discussion. They could be determined from radiological consequence analyses of postulated accidents, as discussed in Section 4.2. For traditional NPPs, an external event-induced core damage frequency smaller that $10^{-5}$ $yr^{-1}$ is considered acceptable in some States.
(8) Potentially risk significant hazards will need a refinement of the preliminary hazard assessment in step (3) above. The other external hazards can keep the preliminary hazard assessment, since their contribution to overall risk is small. This may imply a significant reduction of the effort.

### 4.9.3. Relaxation of qualification requirements

An important outcome of an external event PSA is that SSCs can be ranked by their contribution to risk. In view of this ranking, it may be that the standard plant design specifies an overqualification of some SSCs [39, 41]. For instance, an SSC for which a full seismic qualification is requested, may have an insignificant contribution to seismic risk.

In those cases, the (seismic) fragility of the SSC may be revised, assuming that the SSC is a standard (off the shelf) component, and its risk contribution may be reassessed. If the SSC became a significant risk contributor as result of fragility changes, no relaxation of qualification requirements is possible. However, if the SSC remains a non-significant risk contributor after changing the fragility, then the qualification requirements may be relaxed or even removed.

# 5. APPLICATION OF THE MANAGEMENT SYSTEM TO SITE SURVEY AND SITE EVALUATION PROCESS

The processes siting and site evaluation must be incorporated into the overall management system for the nuclear installation project. The management system needs to be established as early as possible, in line with the implementation schedule for the various activities involved in the site survey and site evaluation stages of the nuclear installation. These activities which have implications relating to nuclear safety are conducted to comply with the Requirement 2 of IAEA SSR-1 [2], which states:

> **"Site evaluation shall be conducted in a comprehensive, systematic, planned and documented manner in accordance with a management system."**

SSR-1 [2] also states:

> "3.1. An integrated management system that meets the requirements of IAEA safety standards Series No. GSR Part 2, Leadership and Management for Safety [42] shall be established. The management system shall cover the organization, planning, work control, verification and documentation of activities and qualification and training of personnel to ensure that the needed quality of the work is achieved at each stage of site evaluation. The management system shall be implemented at the earliest possible time in the conduct of site evaluation for the nuclear installation.
>
> "3.2. Site evaluation shall include, as part of the management system, proper quality assurance arrangements covering each activity that could influence safety or affect the derivation of site specific design parameters and other safety related site characteristics. The quality assurance arrangements shall be consistent with regulatory requirements and their application shall be commensurate with the importance of the activity under consideration to safety."

This clearly indicates that quality assurance arrangements are to be commensurate with the significance of the activity, especially in terms of safety considerations. This is the key for applying a graded approach within the management system and quality assurance programme.

The Safety Guides [3–9] that address siting and site evaluation of various natural and human induced hazards include specific recommendations for the management system to be employed. These recommendations are generally work-oriented and show similarities across different guides, with variations specific to the particular hazard assessment approach being addressed. The common elements for a management system related to site evaluation activities need to be incorporated in a project work plan. Paragraph 10.2 of SSG-9 (Rev. 1) [6] states:

> "A project work plan should be established that, at a minimum, addresses the following topics:
>
> (a) The objectives and scope of the project;
> (b) Applicable regulations and standards;
> (c) Organization of the roles and responsibilities for management of the project;
> (d) Work breakdown, processes and tasks, and schedule and milestones;
> (e) Interfaces among the different tasks (e.g. field tasks, laboratory tests, analysis) and disciplines involved, with all necessary inputs and outputs;
> (f) Project deliverables and reporting."

The methodologies and approaches outlined in lower tier regulations, such as regulatory guidance documents, industry codes and standards, need to be clearly identified and described. If procedures for expert interaction are employed to better address epistemic uncertainties, the complexity and sophistication methods need to be selected based on the specific requirements for the project. The project work plan needs to explicitly detail these approaches and methodologies, including the roles and responsibilities of various experts involved (e.g. proponent, resource expert, technical integrator, review panel member) [6].

In order to demonstrate the traceability and transparency of hazard evaluations to the intended end users such as the operating organization, regulatory body, peer reviewers, designers, vendors, and contractors/subcontractors of the operating organization, all project elements need to be described in the relevant documents for the hazard assessments for both natural and human induced external events. Paragraph 10.8 of SSG-9 (Rev. 1) [6] states:

> "the documentation … should provide a description of all elements of the project, including the following information:
>
> (a) A description of the participants in the evaluation and their roles;
> (b) Background material that contains the analysis documentation, including raw data and processed data;
> (c) A description of the computer software used, and input and output files;
> (d) Reference documents;
> (e) All documents supporting the treatment of uncertainties, expert opinion and related discussions;
> (f) Results of intermediate calculations and sensitivity studies."

Typically, a hazard assessment is carried out to support the design and/or external hazard PSA of a nuclear installation. Consequently, from the outset, the work plan for the hazard assessment needs to clearly define the intended engineering applications and objectives of the assessment, and outline the required outputs — namely, all the results needed to meet the specific engineering purposes and objectives of the assessment [6]. Given the complexity of hazard assessment, a peer review should be incorporated independently into the project work plan to ensure that [6]:

(a) An appropriate and systematic process was followed to conduct the hazard analysis;
(b) The analysis effectively addressed and evaluated the related uncertainties, including both epistemic and aleatory uncertainties;
(c) All the documentations are comprehensive, transparent and traceable.

Some of these work-oriented approaches may be very resource-intensive and time-consuming, and depending on a number of factors a graded approach may be applied.

Requirement 7 of GSR Part 2 [42] states that "**The management system shall be developed and applied using a graded approach.**"

Paragraph 4.15 of GSR Part 2 [42] states:

> "The criteria used to grade the development and application of the management system shall be documented in the management system. The following shall be taken into account:
>
> (a) The safety significance and complexity of the organization, operation of the facility or conduct of the activity;
> (b) The hazards and the magnitude of the potential impacts (risks) associated with the safety, health, environmental, security, quality and economic elements of each facility or activity;
> (c) The possible consequences for safety if a failure or an unanticipated event occurs or if an activity is inadequately planned or improperly carried out."

Taking into consideration the design of some advanced NPPs it may be possible to demonstrate that the above conditions apply. This could be the basis for reducing some of the more onerous tasks related to quality assurance activities. As an example, with respect to the degree of control and verification, factors to be considered for the application of a graded approach may include:

— Extent and form of procedures and instructions;
— Methods of verification (inspection, testing, surveillance and audits);
— Extent of verification (single, redundant or continuous verification methods);

— Need for and extent of audits;
— Amount and content of records.

A graded approach can be used in evaluating different hazards based on the nuclear installation's complexity and potential radiological hazards associated with an intermediate radiological hazard category. Most of the advanced reactors fall in this range. The graded approach can be implemented by either using the same database as used for the NPPs and a lower hazard level corresponding to a higher exceedance probability value or using a simplified method with restricted database and accounting for increased uncertainties due to these two factors.

## Appendix I

## APPLYING A GRADED APPROACH TO PROBABILISTIC SEISMIC HAZARD ANALYSIS FOR A SITE OF INTERMEDIATE OR LOW RADIOLOGICAL HAZARD LOCATED IN A STABLE CONTINENTAL REGION

### I.1. BACKGROUND

Paragraph 9.10 (c) of SSG-9 (Rev. 1) [6] provides two statements regarding the application of a graded approach to vibratory ground motion hazard analysis for installations categorized in an intermediate radiological hazard category:

> "(i) If the seismic hazard assessment is typically performed using methods similar to those described in this Safety Guide, a lower input ground motion than that evaluated for (b) may be adopted for designing these installations, in accordance with the safety requirements for the installation.
>
> (ii) If the database and the methods recommended in this Safety Guide are found to be disproportionately complex, time consuming and demanding for the nuclear installation in question, simplified methods for seismic hazard assessment (that are based on a more restricted data set) may be used. In such cases, the input ground motion finally adopted for designing the installation should be commensurate with the reduced database and the simplification of the methods, with account taken of the fact that both factors tend to increase uncertainties".

In these instances, para. 9.10 of SSG-9 (Rev. 1) [6] states that the input ground motion adopted for design "should be commensurate with the reduced data base and the simplification of the methods, with account taken of the fact that both factors tend to increase uncertainties."

In the probabilistic framework, the first option involves performing a PSHA as for a traditional NPP but allows the use of lower design ground motion levels with higher annual rate of exceedances (e.g. the return period of 5000 years compared to 10 000 years). Therefore, this option does not lead to a reduction in the cost or duration of the PSHA study unless the new facility will be located on a site that already has a PSHA with an appropriate quality assurance pedigree expected for a nuclear installation (e.g. a brownfield site). The second option allows the use of less rigorous processes that use simplified methods or smaller data sets in the hazard assessment, which would result in a reduction of the PSHA study project's cost or duration. In this respect, the second option in applying a graded approach might be more desirable for new nuclear installation projects.

The United States Nuclear Regulatory Commission (NRC) developed the SSHAC procedure to provide a consistent and structured framework for conducting hazard analyses for nuclear installations [43]. Over the last two decades, the SSHAC procedure was utilized in nuclear-related PSHA studies in United States of America and several international projects (e.g. Canada, Japan, South Africa, Spain, Switzerland, Türkiye). The essence of the SSHAC process and the recommendations in SSG-9 (Rev. 1) [6] are similar: the objective is to achieve a well-documented hazard study that captures the centre, body and range of interpretations that are technically defensible. In that sense, the PSHA studies performed for nuclear installations should:

- — Consider the complete range of data, models and methodologies that are potentially relevant to PSHA at the site;
- — Integrate these elements to properly reflect the centre, body and range of interpretations that are technically defensible;
- — Have a thorough documentation and peer review process.

The SSHAC guidelines [43] specified four different levels of SSHAC studies, increasing in complexity and rigor from Level 1 to Level 4. The fundamental objectives of evaluation, integration, documentation and peer-review apply to all levels. However, the different levels have different project structure, and the number of experts included, the duration and cost are highly variable. The NRC expects at least a Level 3 SSHAC study for the PSHA for new NPP site. In this respect, applying a graded approach to vibratory ground motion hazard analysis by using simplified methods based on a more limited data range as given in SSG-9 (Rev. 1) [6] would be analogous to moving from Level 3 to a lower level SSHAC study for nuclear installations.

Practical experience from recent applications of the SSHAC procedure shows that changing the efforts of a Level 3 PSHA to a Level 1 PSHA may reduce the project duration by a factor of three (three or more years to one year) and may reduce the budget of the project up to a factor of ten [18]. The reduction in time requirements may be very significant for new builds in intermediate radiological hazard category. Feedback from past practice showed that the largest increments of time, effort, complexity and cost occur in passing from Level 2 to Level 3. Therefore, performing a SSHAC Level 2 PSHA for nuclear installations in the intermediate radiological hazard category may result in the appropriate level of rigor and regulatory assurance expected for a nuclear installation PSHA, and with reasonable project cost and duration.

The necessary effort in data collection and hazard modelling during the integration phase of the PSHA is correlated with the evaluation of epistemic uncertainties [34]. With a reduced effort and models with greater simplification selected, it is likely that the epistemic uncertainty (i.e. range of the hazard estimates about the central or median estimates) is underestimated [6]. It is necessary that a graded PSHA study account for the underestimation in the epistemic uncertainty. If sufficient margin exists between the design level of the facility and the hazard estimate, the increase in the epistemic uncertainty may be accommodated by the design. Therefore, applying a graded approach to the PSHA efforts by using simplified methods or a more restricted data set is more likely to be practical in regions with relatively low seismic activity.

The primary aim of this publication is to present a practical approach for applying a graded approach to the PSHA efforts for new sites of nuclear installations within the intermediate radiological hazard category. The proposed method is applicable to the sites located in stable continental regions with relatively low seismicity, including, but not limited to, Australia, Canada, northern Europe, portions of South America, central and eastern parts of United States of America, and West Africa. For installation sites located in shallow crustal active tectonic regions and subductions zones, site specific conditions (proximity to major tectonic structures) may a have a significant impact on the PSHA result. Therefore, a sufficient margin may not exist to justify the application of a graded approach based on the cost of increased uncertainty. A case-by-case evaluation is necessary for sites located in high seismicity areas.

#### I.1.1. Contribution of Member States experience on proposed methodology

This publication uses the significant experience accumulated by Member States in performing PSHAs for nuclear installations for proposing the practical application of a graded approach to PSHA in two different ways. First, the site specific SSHAC Level 2–4 PSHAs performed since 2011 (after the accident at the Fukushima Daiichi NPP) were collected to compare the site specific seismic hazard estimates for nuclear installation sites with publicly available hazard curves that are used in national and regional seismic hazard maps. Second, the most recent SSHAC Level 3 PSHA efforts were analysed to estimate the range of epistemic uncertainty implemented in these projects, in terms of the ratio of the hazard fractiles (e.g. mean to median, mean to 84th percentile). Both analyses aim to provide scaling factors that will be used in the overall procedure to get a conservative hazard estimate.

### I.2. PROPOSED FRAMEWORK FOR A GRADED APPROACH TO PROBABLISTIC SEISMIC HAZARD ANALYSIS EFFORTS IN STABLE CONTINENTAL REGIONS

The flow chart presented in Fig. 6 summarizes the proposed framework for applying a graded approach to PSHA efforts for nuclear installations within the intermediate radiological hazard category in stable continental regions (SCRs). Each step of the procedure is explained in detail in the following subsections.

### I.2.1. Compliance of probabilistic seismic hazard analysis performed for national or regional seismic hazard maps (Step 1)

The seismic hazard maps developed at national or regional levels provide a benchmark for the site specific PSHA results if these hazard maps were developed based on the objectives that are relevant for the PSHA of nuclear installations. Specifically, the national or regional seismic hazard map needs to: (a) consider the complete set of national and international data, models and methodologies available at the time of their development, (b) include a logic tree structure that captures the centre, body and range of interpretations that are technically defensible and (c) have a clear and publicly accessible documentation for the seismic source model and ground motion characterization model. Therefore, the first step of the proposed procedure is to review the available national or regional seismic hazard maps to answer the following questions:

(1) Do the PSHA calculations for the hazard map include and properly model the epistemic uncertainty in both the seismic source model and ground motion characterization model?
(2) Is the distribution of the ground motion models' aleatory variability truncated properly?

Epistemic uncertainty in the current seismic source characterization models is typically represented by complex logic trees developed for several seismic source parameters. Due to the computational demands to develop high resolution hazard maps, these complex logic trees are generally trimmed in PSHA calculations for large regions. In this regard, the logic tree implemented in the PSHA used for developing the map needs to at least include:

— Alternative correlated pairs of the parameters of magnitude-frequency models (the slope $b$ and the intercept $a$ as defined in para. 4.30 of SSG-9 (Rev. 1) [6]);
— Alternative potential maximum magnitude values for each seismic source, represented by either discrete or continuous functional form (see para. 4.26 of SSG-9 (Rev. 1) [6]);
— For fault sources, alternative values of maximum rupture dimensions (see para. 4.14 of SSG-9 (Rev. 1) [6]) and the rate of earthquake activity such as slip rate (see para. 4.19 of SSG-9 (Rev. 1) [6]).

In addition:

— The thickness and the depth of the seismic source zones should be characterized properly (see para. 4.22 of SSG-9 (Rev. 1) [6]);
— The size of the seismic source zones of homogeneous spatial distribution for activity rates need to be large enough to provide statistically stable estimates of the magnitude–frequency model parameters and small enough to consider the changes in the tectonic characteristics or spatial variability in the activity rate.

Figure 7 (left panel) provides an example from recently developed European seismic hazard map [44], showing that the size of the seismic source zones varies according to the data availability and source kinematics. The logic tree for seismic source zones includes alternatives for the magnitude-frequency distribution parameter pairs (a, b) and maximum magnitude. Figure 7 (right panel) shows a seismic source zone adopted in the 2018 version of US national hazard maps and the logic tree branches for earthquake recurrence and maximum-magnitude models.

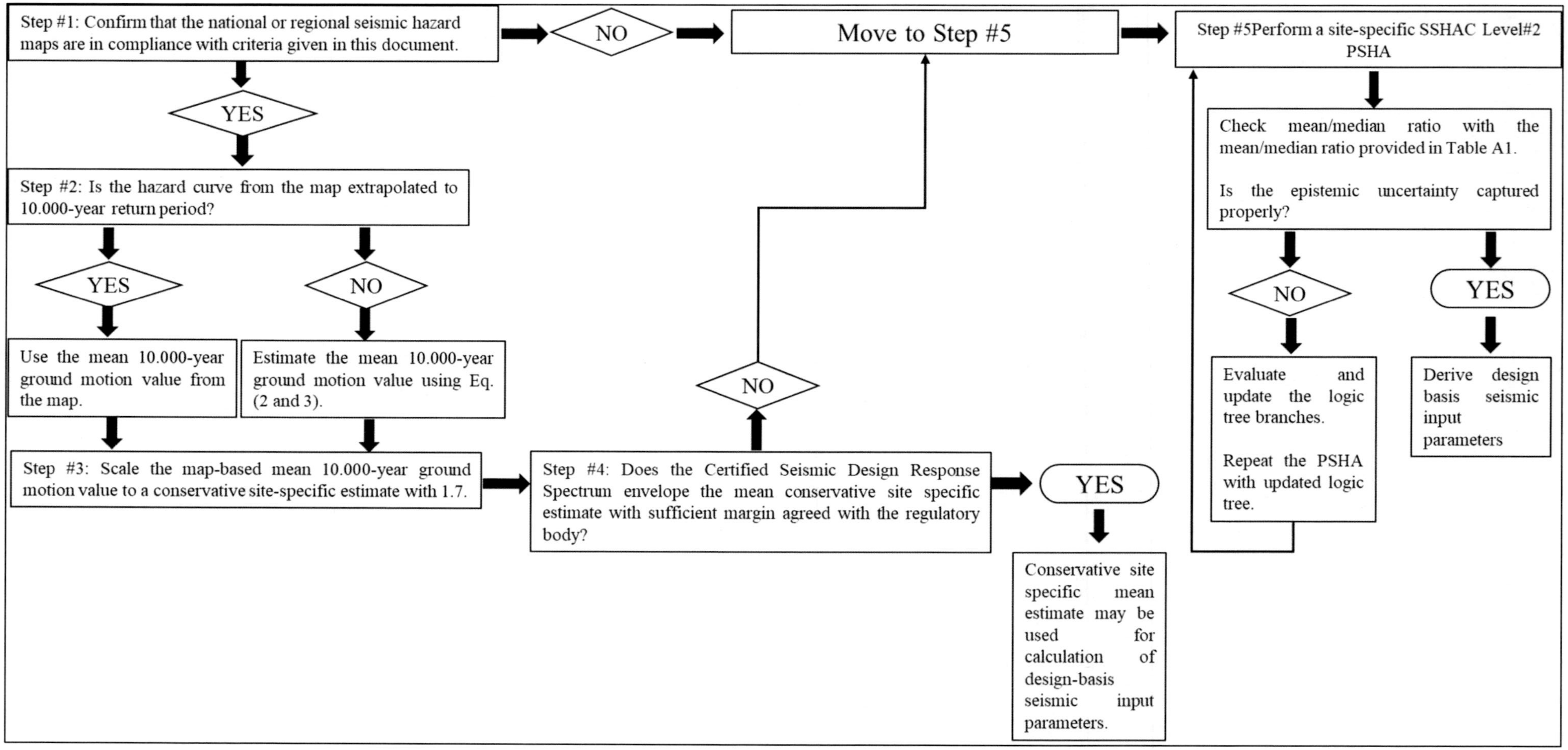


*FIG. 6. Flow chart of the proposed graded approach for nuclear installations within the intermediate and low radiological hazard category in stable continental regions.*

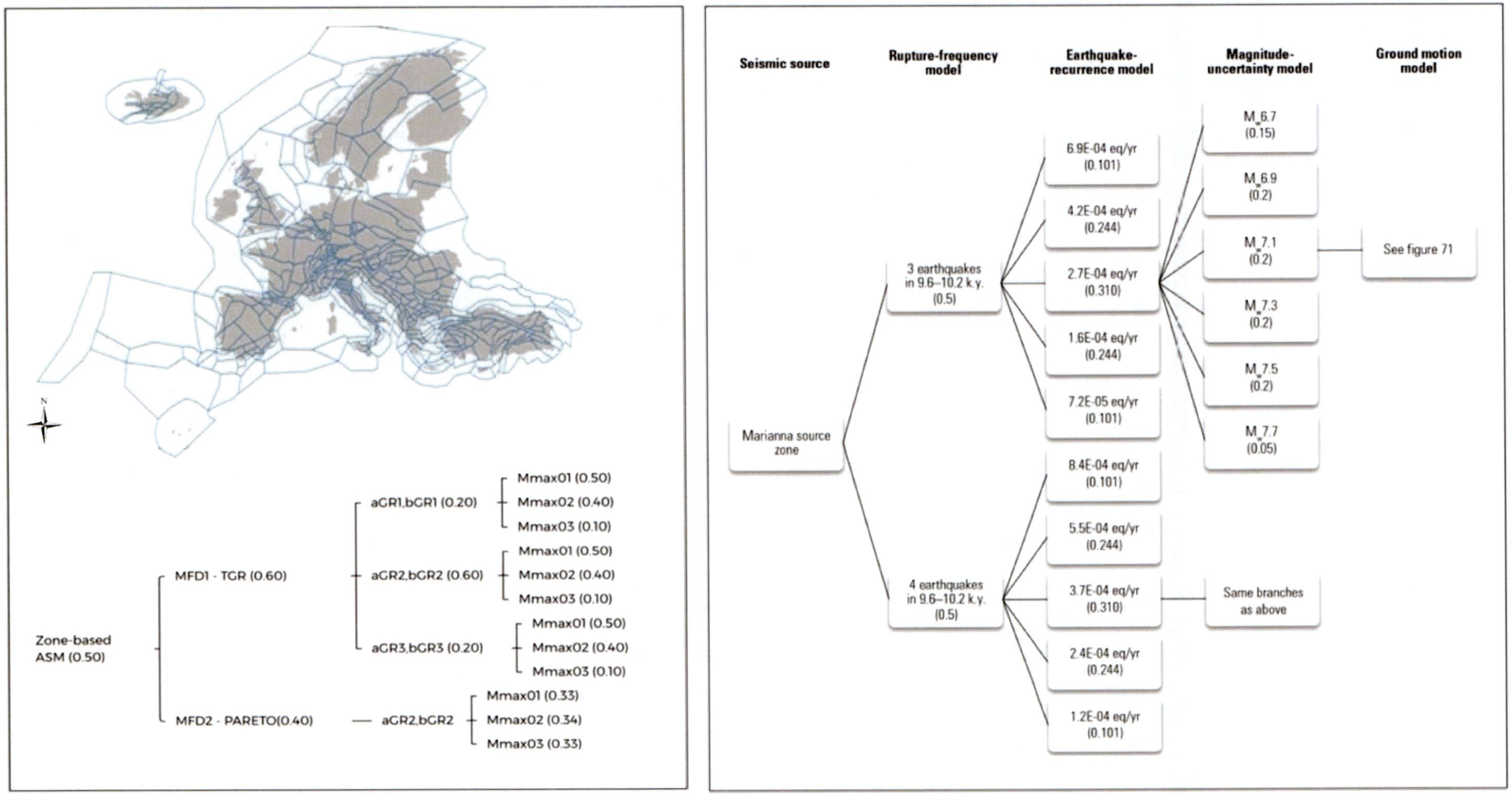


*FIG. 7. Left panel: seismic source zones and associated logic tree for seismic source characterization model in ESHM 2020 (adapted from Ref. [44]). Right panel: an example source characterization logic tree used in 2018 US national hazard maps (reproduced from Ref. [45]).*

The ground motion models' aleatory variability is represented by the lognormal distribution of ground motion residuals. In current PSHA studies, this distribution may be modelled with a truncation at two or three standard deviations or modelled without any truncation as in 2018 US national hazard maps [45]. Figure 8 compares the hazard curves generated for a low seismicity site with different values of truncation as also discussed in Ref. [46]. For design basis ground motions return period (e.g. an annual exceedance probability of $10^{-4}$), the truncation has a negligible effect for truncation at more than three standard deviations. Therefore, national or regional hazard maps should be checked for truncation of ground motion variability (if applied) to ensure that the truncation is not less than three standard deviations.

Based on the above considerations, most recent national hazard maps of Canada and the United States of America, as well as the 2020 European seismic hazard map [44] may be used as a benchmark for the mean conservative site specific estimate for design basis ground motions of nuclear installations within the intermediate radiological hazard category in SCRs. As these hazard maps provide the mean hazard curves up to a 10 000 year return period, an additional step is not necessary to get the map-based ground motions associated with the mean hazard.

New information on the seismic sources may be available after the development phase of the seismic source model of the national seismic hazard map (e.g. a new fault has been mapped or an existing fault may be found to be capable). If this new information is related to seismic sources near the site of interest, the site specific ground motions could differ significantly from the ground motion estimates provided by the national hazard map. Therefore, even if the national seismic hazard map complies with the criteria given above, a review of the seismic source model of the map is needed for updates in the current literature.

### I.2.2. Extrapolating the hazard curves to a 10 000 year return period (Step 2)

If the national or regional hazard map selected as a benchmark does not provide the hazard estimates for long return periods, the 10 000 year return period's ground motions may be estimated by extrapolating the hazard curve using the ground motions with the longest return period that is provided by the hazard map. Typically, the hazard maps provide peak ground acceleration (PGA) and pseudo-spectral accelerations (Sa) at several spectral periods at two hazard levels, e.g. for 475 year return period and 2475 year return period.

Given that the PGA corresponding to two hazard levels, $\mathbf{Haz_1}$ and $\mathbf{Haz_2}$ are taken from the map, the PGA for the 10 000 year return period may be computed using linear interpolation on the natural logarithm of annual frequency of exceedance and PGA as shown in Eq. (3).

$$\ln(\text{PGA(10,000-yr)}) = \frac{1}{k}\ln\left(\frac{0.0001}{\text{Haz}_1}\right) + \ln\left(\text{PGA}(\text{Haz}_1)\right) \quad (3)$$

The slope of the log-hazard curve, k, is calculated by using Eq. (4):

$$k = \frac{\ln\left(\frac{\text{Haz}_2}{\text{Haz}_1}\right)}{\ln\left(\frac{\text{PGA}(\text{Haz}_2)}{\text{PGA}(\text{Haz}_1)}\right)} \quad (4)$$

If the benchmark seismic hazard maps provide the ground motions for only one return period, the hazard curve's slope can be conservatively approximated by $k = -1.5$ for SCRs (Fig. 8). If needed, Eq. (3) may be modified to estimate the ground motion intensity measures with different return periods.

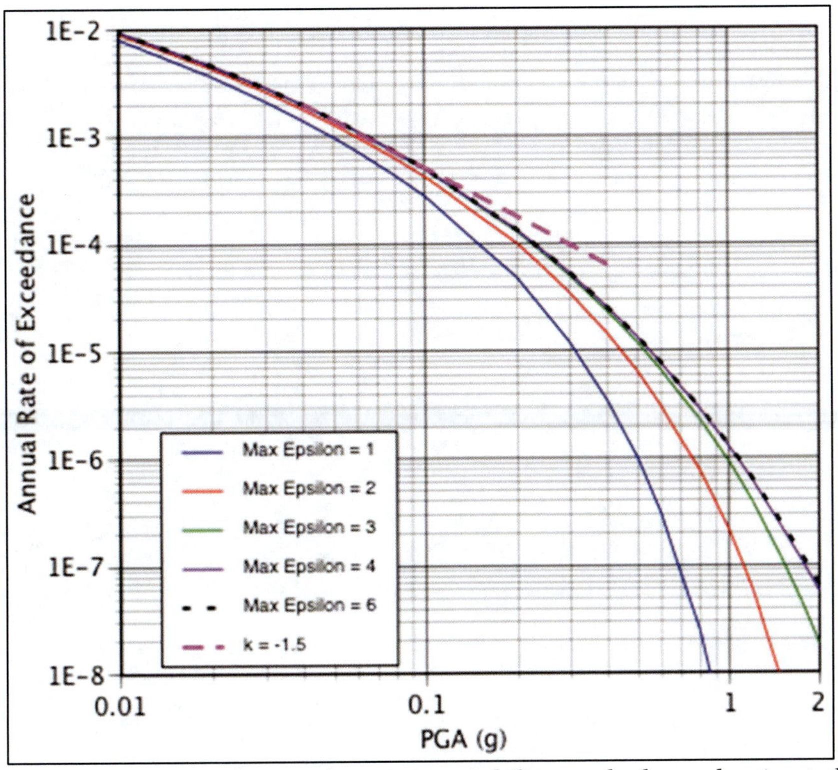


*FIG. 8. Effect of the truncation of ground motion variability on the hazard estimated for a site with low seismicity. Using k= −1.5 provides a conservative slope to extrapolate the hazard results available in national maps to longer return periods.*

### I.2.3. Conservative site specific estimate of the ground motion for the mean hazard (Step 3)

Site specific PSHA for nuclear installations is benefited by the use of complex and well-established logic trees, developed with the support of a comprehensive and integrated database that include geological, geophysical, geotechnical and seismological data acquired for the project. Collection of sufficient reliable and relevant site specific data will reduce the epistemic uncertainty; however, the mean site specific hazard may be lower or higher than the mean hazard estimate provided in the benchmark seismic hazard map. To evaluate the difference between the site specific and map-based hazard estimates, data from 62 NPP sites was compiled as shown in Figure 9. These sites are selected based on the following criteria: (a) they are located in (or close to) stable continental regions, (b) a SSHAC Level 2–4 PSHA was performed for these sites in the last decade and, (c) the site specific PSHA results are available to the authors. The map-based 10 000 year PGA values were compiled either from 2018 US national hazard map or 2020 European seismic hazard map for rock site conditions ($V_{S30}$ = 760–800 m/s). As the hazard estimates change significantly with site conditions, the hazard curves for rock site conditions are taken from site specific PSHA reports as much as possible. Therefore, the site specific PGA values used in this study are not equal to the design basis ground motions at these particular NPP sites.

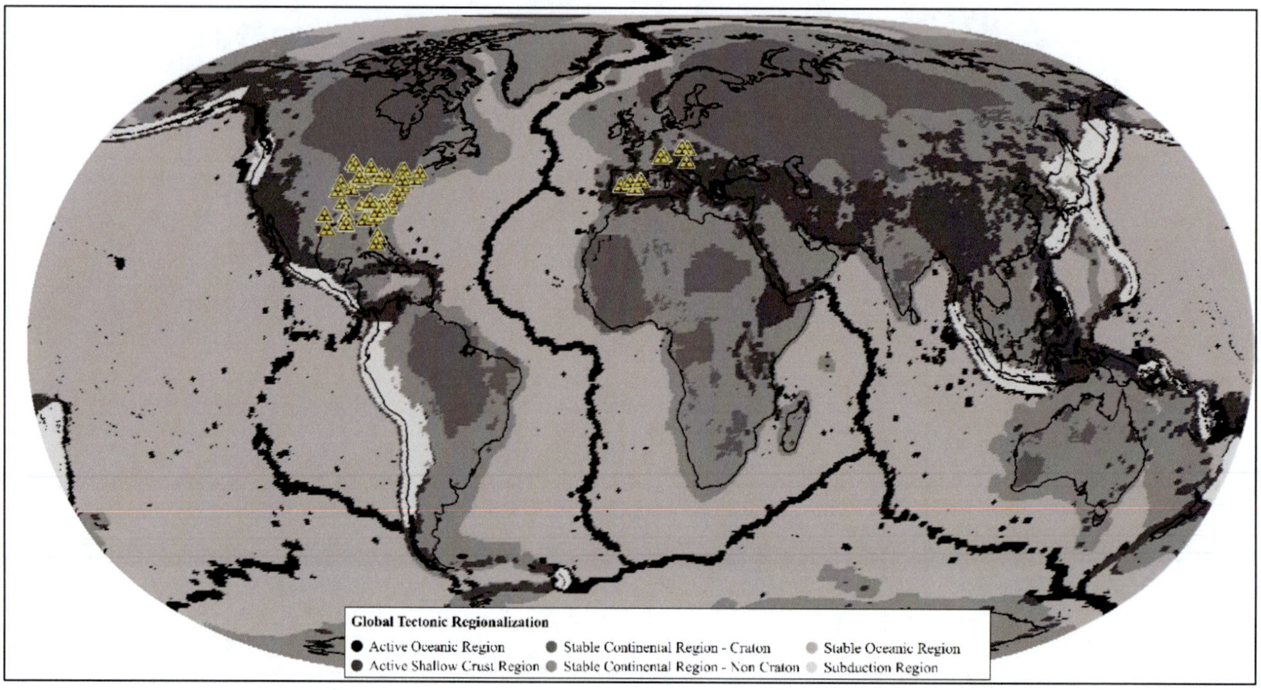


*FIG. 9. The NPP sites used to calculate the map-based to site specific ratio of 10 000 year return period PGA, overlaid by a tectonic categorization of the globe.*

Figure 10 presents the distribution of the ratio of site specific to map-based PGA with 10 000 year return period. In 66% of the collected cases (41 sites), the site specific PGA value is lower than the map-based PGA value for the 10 000 year return period. Mean, median, and standard deviation (σ) of the ratios is 1.04, 0.88 and 0.66, respectively, indicating that the distribution is skewed to left and the map-based values provide an overestimation for site specific values for most of the cases. The ratio of site specific to map-based PGA lies between 1.1 and 1.7 (mean+1σ range) for 18 NPP sites (29%). For seven sites, the site specific PGA value is larger than 1.7 times the map-based value. As the objective is to find a scale factor that can be applied to the map-based values that covers the range for most sites, the map-based PGA value for a 10 000 year return period may be scaled by 1.7 to develop the conservative estimate of the site specific PGA with 10 000 year return period.

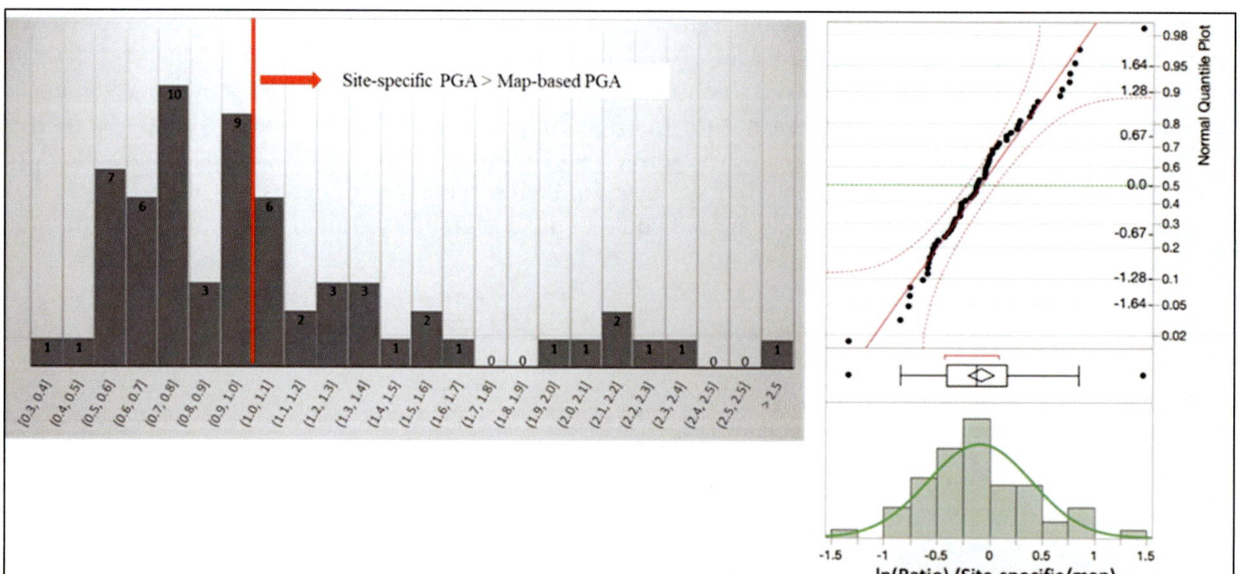


*FIG. 10. Site specific to map-based ratio of 10 000 year return period PGA values for selected NPPs.*

### I.2.4. Comparing the conservative site specific estimate of the ground motion for the mean hazard with the certified seismic design response spectrum (Step 4)

At this step, the conservative estimate of the site specific ground motions for the mean hazard with a 10 000 year return period will be compared with the certified seismic design response spectrum. For that comparison, in addition to PGA, a full response spectrum may be needed. In that case, the design spectral shape provided by the national design code combined by the conservative estimate of the PGA with 10 000 year return period would be used as a conservative estimate of the site specific design spectrum. If the certified seismic design response spectrum conservatively envelopes the estimate of the site specific mean spectrum, then the conservative estimate may be used.

### I.2.5. Performing the SSHAC Level 2 probabilistic seismic hazard analysis (Step 5)

If the certified design spectrum does not envelop the conservative estimate of the site specific spectrum, or if national/regional seismic hazard maps cannot provide a benchmark value for the site specific ground motions, then a SSHAC Level 2 PSHA should be performed, as recommended at step 5 in the flow chart (Fig. 6). The characterization of seismic source and ground motion logic trees for the SSHAC Level 2 PSHA may need to be developed based on a smaller data set, especially if the site is not yet fully characterized. Therefore, the epistemic uncertainty's range involved in the overall logic tree structure needs to be evaluated by checking the range of epistemic uncertainty in hazard fractiles. A baseline estimate for the appropriate epistemic uncertainty range may be developed based on the previous SSHAC projects.

For this purpose, a total of 28 sites with a SSHAC Level 3 PSHA, 23 from the central and eastern United States of America (CEUS) sites and five from Spain, are analysed for the range of epistemic uncertainty in hazard fractiles. For the 23 CEUS sites, the seismic hazard was recently revised through resolution of Generic Issue 199, which the NRC initiated in 2005, utilizing the SSHAC Level 3 protocol [47]. The seismic hazard reassessment for five NPPs in Spain was conducted from 2015 to 2022, also employing a SSHAC Level 3 protocol. Table 4 presents the mean and standard deviation for the mean and median ratios for 26 sites, for spectral frequencies of 0.5, 1.0, 2.5, 5.0, 10.0, 25.0 and 100.0 Hz, and annual frequencies of exceedance of $10^{-4}$, $10^{-5}$ and $10^{-6}$ years. Figure 11 shows the distribution of mean/median ratios obtained for all those sites [48].

The ratio of mean to median remains close to a constant value or within a relatively narrow range, particularly for spectral frequencies between 2.5 Hz and 10 Hz. The 2.5–10 Hz frequency range encompasses the most critical frequencies for nuclear installations' seismic design. For a 10 000 year return period, the mean to median ratio of 1.35 would envelop most of the investigated sites for the 2.5–10 Hz frequency range. For spectral frequencies between 25 and 100 Hz, a slightly larger mean to median ratio (1.55) is obtained using the mean to median ratio with one standard deviation. A mean to median ratio larger than 1.6 is not obtained for any of the cases for the 10 000 year ground motions. To improve the statistical significance of the data presented in Table 4, it is essential to expand the database for sites with available SSHAC Level 3 PSHA results as much as possible. This might be difficult due to confidentiality of the PSHA results in some States; however, with the current level of data availability, the mean to median ratio of 1.6 may be used as a conservative estimate of the range of epistemic uncertainty to be involved in a SSHAC Level 2 PSHA study in SCRs.

TABLE 4. STATISTICS MEAN/MEDIAN SPECTRAL ACCELERATIONS RATIOS DERIVED FROM HAZARD CURVES IN 26 PROBABILISTIC SEISMIC HAZARD ANALYSES CONDUCTED USING THE SSHAC LEVEL 3 PROTOCOL

| Spectral frequency (Hz) | 10 000 year return period | | 100 000 year return period | | 1 000 000 year return period | |
|---|---|---|---|---|---|---|
| | Mean of the (mean/median ratio) | Standard deviation of the (mean/median ratio) | Mean of the (mean/median ratio) | Standard deviation of the (mean/median ratio) | Mean of the (mean/median ratio) | Standard deviation of the (mean/median ratio) |
| 0.50 | 1.40 | 0.12 | 1.51 | 0.15 | 1.64 | 0.13 |
| 1.00 | 1.29 | 0.10 | 1.33 | 0.08 | 1.39 | 0.08 |
| 2.50 | 1.22 | 0.12 | 1.24 | 0.12 | 1.28 | 0.09 |
| 5.00 | 1.19 | 0.08 | 1.20 | 0.08 | 1.22 | 0.08 |
| 10.0 | 1.21 | 0.08 | 1.20 | 0.08 | 1.21 | 0.10 |
| 25.0 | 1.31 | 0.17 | 1.27 | 0.18 | 1.28 | 0.20 |
| 100.0 | 1.35 | 0.10 | 1.30 | 0.12 | 1.30 | 0.11 |

Note: Statistics shown in the table correspond to 21 CEUS sites and five sites in Spain.

If a SSHAC Level 2 PSHA study is conducted, the full uncertainty range may not be captured. The median hazard is controlled by the central branches of the logic tree, whereas the mean hazard can be sensitive to the weights given to the upper branches of the logic tree. For a SSHAC Level 2 PSHA, there is more confidence in the median than the mean hazard. As a check on the uncertainty for the SSHAC Level 2 study, the ground motion's ratio for a 10 000 year return period between the mean hazard and the median hazard needs to be evaluated. If the ratio is less than 1.6, the epistemic uncertainty may be underestimated. In this case, the ground motion for the mean hazard needs to be estimated by scaling the ground motion from the median hazard by a factor of 1.6.

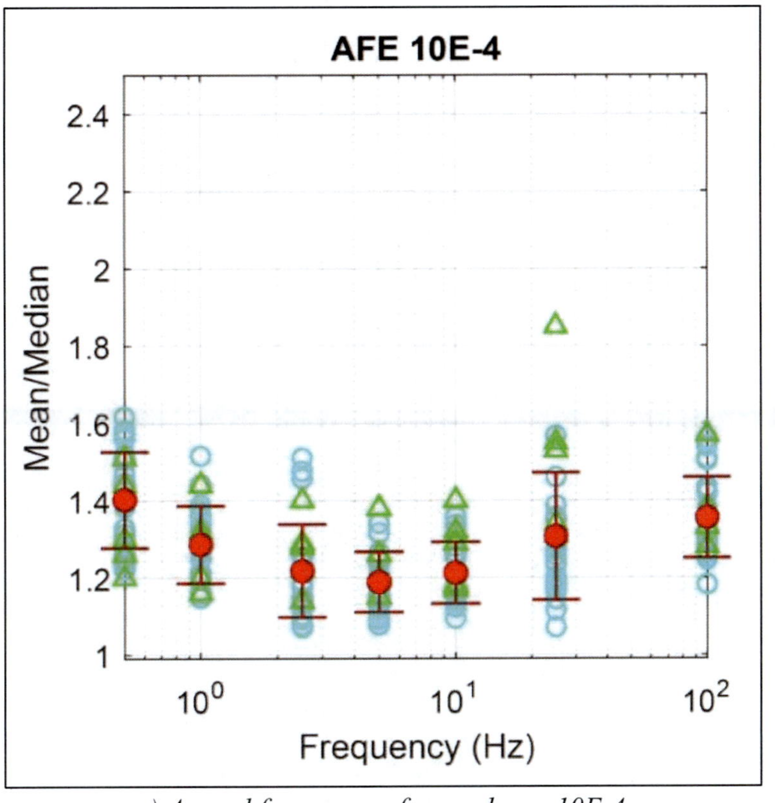


*a) Annual frequency of exceedance 10E-4.*

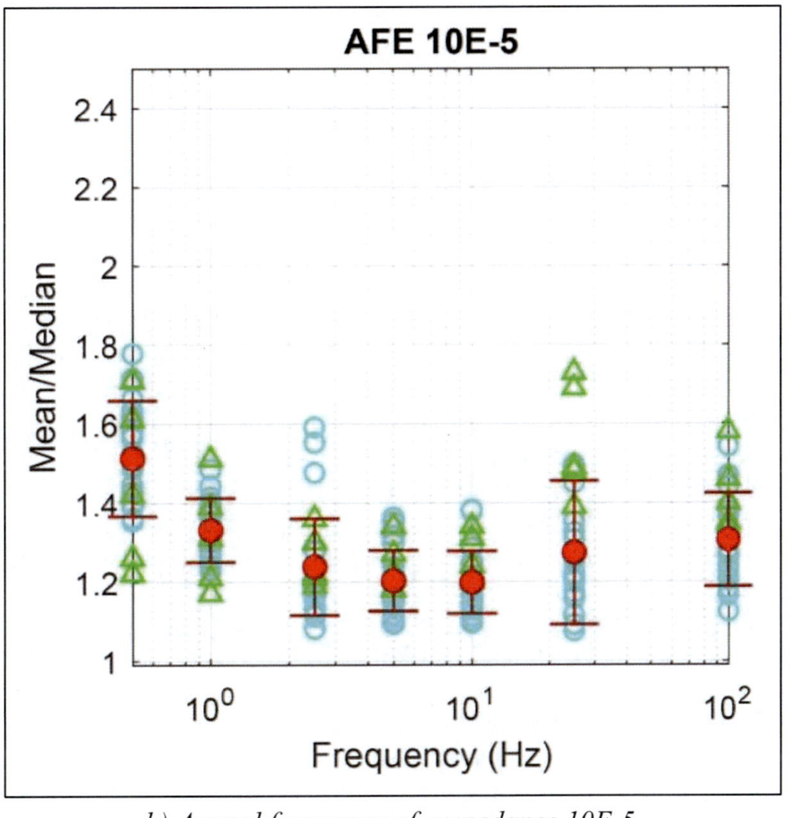


*b) Annual frequency of exceedance 10E-5.*

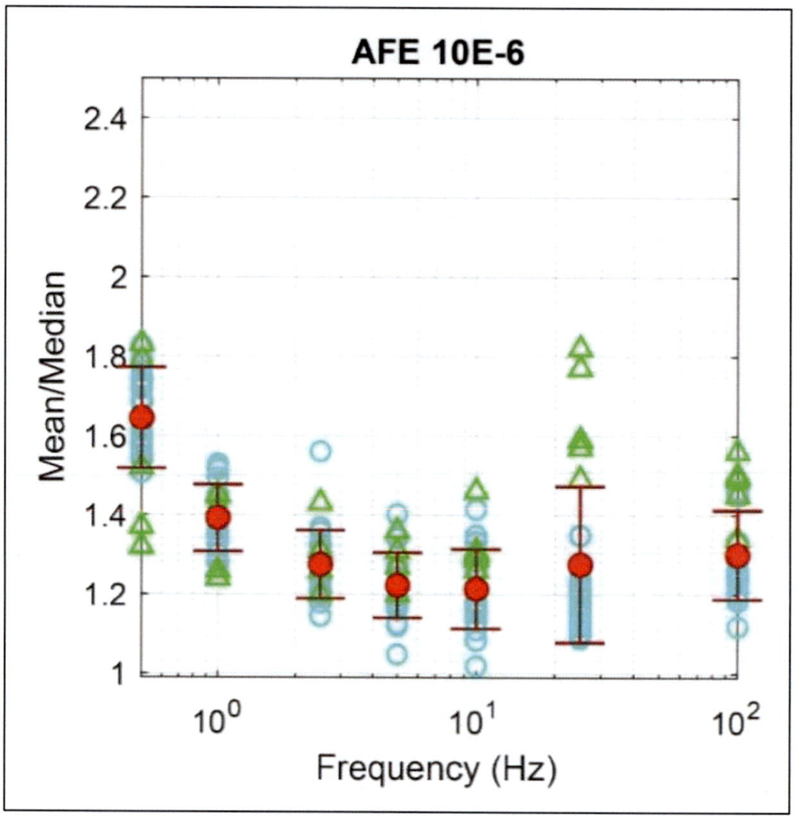


*c) Annual frequency of exceedance 10E-6.*

*FIG. 11. The seismic hazard results from 26 SSHAC Level 3 studies showing mean/median spectral acceleration ratios. Solid red dots indicate the mean values of the ratios, with red error bars representing the range of mean +/– one standard deviations. Blue circles correspond to CEUS sites, while green triangles represent sites in Spain.*

## I.3. CONSERVATISM INCLUDED IN THE PROPOSED GRADED APPROACH

Steps 1–4 of the proposed procedure for applying a graded approach allow the user to have a crude but conservative estimate of the vibratory ground motion hazard for intermediate and low hazard category nuclear installations. This crude, but conservative, estimate of the hazard may also be used for site or technology selection for installations categorized in an intermediate radiological hazard category. A crude estimate of the hazard may also be achieved by performing a baseline PSHA using available data and models. Simplifications in the baseline PSHA project may include the elimination of broad or distant sources or faults from further study that do not contribute to the hazard, using a single (zoneless) areal source zone with spatial smoothing of seismicity rates, using a backbone ground motion model, or elimination of the low-weight branches from the logic tree. On the other hand, the scope of a baseline PSHA is not as well defined. As a result, it may be challenging to review and assess the baseline PSHA, and the expected reduction in duration and cost of the project may not be achieved. In addition, the baseline PSHA may underestimate the epistemic uncertainty.

Estimating the conservative mean hazard from a well-established national or regional seismic hazard map is faster and better constrained compared to the estimates of an unsupervised baseline PSHA study. In many parts of the world, the national or regional seismic hazard maps have integrated all available data and models, are well-documented, open source and peer reviewed. As the site specific components of a PSHA are excluded in the mean estimate from the hazard map, the mean 10 000 year ground motions from the map are multiplied by 1.7 for additional conservatism (Step 3) to reflect the limitations of the regional study as compared to a site specific study. This conservatively biased mean hazard estimate may be acceptable for SCRs, as the standard design of the nuclear installation in the intermediate radiological hazard category is expected to be significantly higher.

Noting that the geotechnical site parameters may be unknown at this stage, the conservative estimate is provided for rock-site conditions. If the shear wave velocity profile of the site is available, a simple site response analysis may be conducted to estimate the site amplification factors.

If a well-established national or regional seismic hazard map is not available, a full scope PSHA is needed for the determination of seismic design input parameters and safety evaluation of the nuclear installation in intermediate or low hazard categories. A SSHAC Level 2 PSHA is recommended for this step (Step 5). Using results at the site characterization stage, site parameters (e.g. site stiffness) will be incorporated in the SSHAC Level 2 PSHA, which will lead to a more realistic vibratory ground motion hazard. The proposed procedure also includes an estimate of the epistemic uncertainty to be included in a SSHAC Level 2 PSHA in SCRs that may have reliable median hazard values but underestimate the epistemic uncertainty.

For high seismicity sites, performing a SSHAC Level 3 PSHA may be unavoidable, because the conservatively biased hazard estimate may significantly exceed the certified seismic design response spectrum.

The correction factors proposed in Section I.2.5 compensate for the reduction of incorporated uncertainties, which results from the reduced SSHAC level of the PSHA.

## Appendix II

## APPLICATION OF A GRADED APPROACH TO GEOTECHNICAL INVESTIGATION AT NUCLEAR INSTALLATION SITES

### I.4. A PERFORMANCE BASED GRADED APPROACH TO SITE GEOTECHNICAL INVESTIGATION

#### I.4.1. Background on structural reliability concepts

In a simple problem, in which structural or geotechnical capacity can be characterized by a single variable R, and demand can be defined by a single variable S, failure occurs when $R-S < 0$.

In general, *R* and *S* will be random variables, with mean $\mu$ and standard deviation $\sigma$. Probability of failure $p_f$ is given by the probability that $M = R-S < 0$ (shaded area in Fig. 12). The probability of failure $p_f$ can be obtained from the probability distributions of R and S [49].

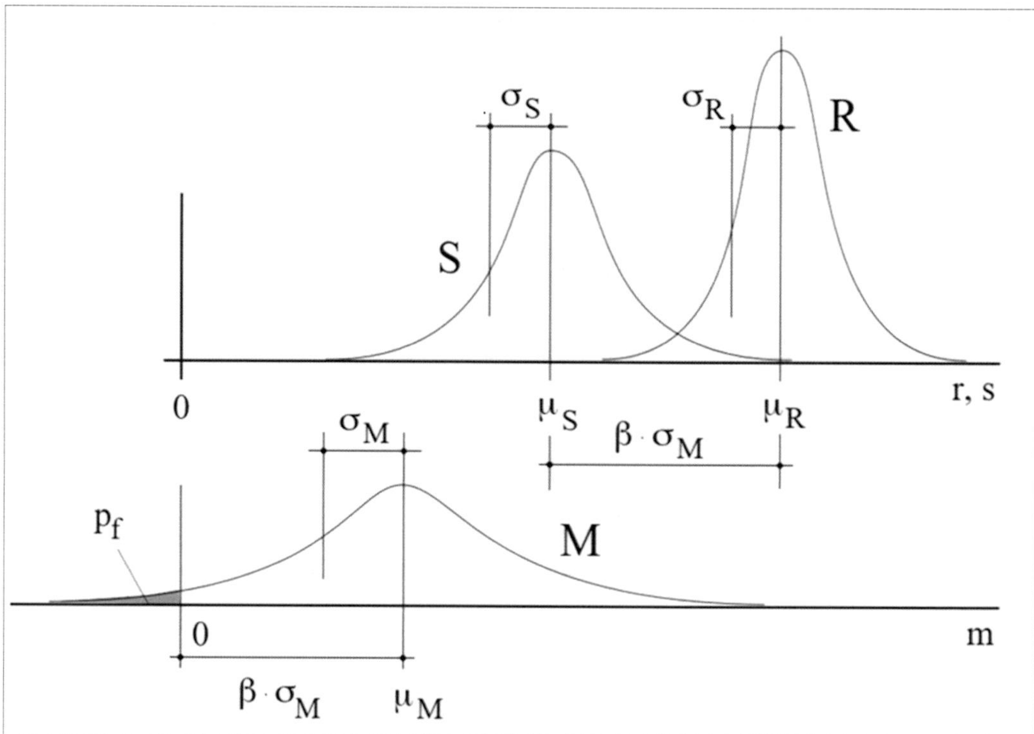


*FIG. 12. Sample set of seismic hazard curves (reproduced from Ref. [49]).*

Geotechnical investigation campaigns may be interpreted as sampling problems. For each geotechnical parameter, there is a probability distribution characterized by a true mean $\mu_{true}$ and a true standard deviation $\sigma_{true}$. The latter corresponds to aleatory variability, and it cannot be reduced, even with infinite boreholes or tests. Sometimes, this is termed as physical uncertainty about the parameter (see Ref. [50]).

Ideally, the true mean $\mu_{true}$ and the true standard deviation $\sigma_{true}$ would be used to derive design basis geotechnical parameters, with the purpose of achieving a probability of failure $p_f$ smaller than a target value. Note that, in many geotechnical problems, it is the demand on the ground S (e.g. vertical pressure on the soil) which is controlled by the designer, rather than the capacity R (e.g. ultimate bearing capacity).

In practice, only a limited number of boreholes, explorations and tests are conducted, from which approximate values of the mean and standard deviation could be derived: $\mu_{approx}$ and $\sigma_{approx}$.

Classical sampling theory provides confidence limits for the true mean $\mu_{true}$ and the true standard deviation

$\sigma_{true}$., computed from mean and standard deviation of the sample and the number of observations (e.g. tests) in the sample [36]. For the same confidence level (e.g. 95%), the larger the size of the sample (i.e. the number of observations), the smaller the confidence intervals are. For a very large number of observations, the sample mean tends to the true mean and standard deviation of the observations tends to the aleatory variability of the parameter. For smaller samples, the confidence limits can be interpreted as defining the range of epistemic uncertainty, which refers to the uncertainty that can be lowered by increasing the observation numbers. In the literature, this epistemic uncertainty sometimes has been designated as statistical uncertainty.

### I.4.2. Geotechnical investigation campaigns for traditional nuclear power plants

Traditionally, geotechnical investigation campaigns for NPPs were intended to obtain a prudent estimate of the true mean $\mu_{true}$ of the relevant geotechnical parameters. For design purposes, the mean value was adjusted by a global safety coefficient, based on experience and judgement, to account for uncertainties.

The global safety coefficient method is slowly being abandoned by States in favour of the so-called partial safety coefficients method, which considers uncertainty in the demand $S$ separately from the uncertainty in the capacity $R$, as depicted in Fig. 12. This method allows computation of the probability of failure $p_f$ (performance) of a foundation or geotechnical structure.

At traditional NPPs, geotechnical investigation campaigns are now intended to obtain the real mean $\mu_{real}$ and the real standard deviation $\sigma_{true}$ of ground parameters. Hence, a large number of observations are normally made. From $\mu_{true}$ and $\sigma_{true}$, a high confidence or characteristic value of the corresponding geotechnical parameter is obtained, which is then used in the design calculations.

From a real NPP site, where a very large number of observations were made to determine geotechnical parameters, Table 5 shows the mean and standard deviation values obtained at different locations.

In this case, mean and standard deviation of the sample are good approximations to the true values and can be used to obtain high confidence values of the parameters.

TABLE 5. STATISTICS OF DIFFERENT GEOTECHNICAL PARAMETERS DERIVED FROM A VERY LARGE NUMBER OF OBSERVATIONS AT A SPECIFIC SITE

| Geotechnical parameter | Location | Mean | Std. deviation | Coef. of variation |
|---|---|---|---|---|
| Specific weight (ton/m$^3$) | Layer 2 | 1.96 | 0.03279 | 0.0167 |
| | Layer 3 | 2.05 | 0.01639 | 0.0080 |
| Cohesion (kPa) | Layer A | 63.39 | 3.27869 | 0.0517 |
| | Layer B | 694.23 | 85.2459 | 0.1228 |
| | Layer C | 846.64 | 180.3279 | 0.2130 |
| Ultimate strength (MPa) | Layer V | 5.86 | 0.3770 | 0.0643 |
| | Layer N | 14.57 | 2.7377 | 0.1879 |
| | Layer Y | 2.40 | 0.1148 | 0.0478 |

### I.4.3. Application of a graded approach to the site investigation campaign

As mentioned above, sampling theory can be used to obtain confidence intervals for the true mean $\mu_{true}$ and the true standard deviation $\sigma_{true}$ of the statistical distribution of a geotechnical parameter. If the parameter has a normal (Gaussian) statistical distribution, Table 6 provides confidence limits for 95% confidence, as a function of the number $n$ in the sample's observations and coefficient of variation (COV) of the sample.

TABLE 6. CONFIDENCE LIMITS AS A FUNCTION OF THE NUMBER OF OBSERVATIONS AND THE COEFFICIENT OF VARIATION OF THE SAMPLE [36]

| Observations (n) | Sample COV | Level of confidence (%) | Confidence limit of (sample mean–true mean)/true mean | Confidence limit of true COV |
|---|---|---|---|---|
| 6 | 0.15 | 95 | 0.1188 | 0.343 |
| 10 | 0.15 | 95 | 0.0859 | 0.260 |
| 12 | 0.15 | 95 | 0.0771 | 0.243 |
| 20 | 0.15 | 95 | 0.0577 | 0.211 |
| 40 | 0.15 | 95 | 0.0398 | 0.184 |
| 60 | 0.15 | 95 | 0.0323 | 0.177 |

The relative error of the sample mean is reduced by increasing the number of observations $n$. The reduction as a function of $n$ is very quick at the beginning and it progressively slows down. The same happens with the confidence limit of the true COV. In view of the data in Table 6, going beyond 20 observations is very inefficient and is likely not warranted, given the necessary effort.

In a real case, the scope of the geotechnical characterization campaign will be very site dependent and will need to be defined by an experienced team of qualified professionals. However, looking for some general guidance on how to apply a graded approach to the effort for nuclear installations that are less hazardous than traditional NPPs, one may question how a reduction of the effort in characterization may influence the performance (probability of failure $p_f$) of a foundation or geotechnical structure.

In a simple problem, in which structural or geotechnical capacity can be characterized by single variable $R$, and demand can be defined by a single variable $S$ (see Fig. 12), the probability of failure $p_f$ can be obtained relatively easily if the distributions of $R$ and $S$ are assumed to be normal [49].

It is reasonable to assume that the scope of geotechnical investigation for a traditional NPP will lead to an acceptable performance or probability of failure $p_f$. If the scope is reduced in accordance with a graded approach, then uncertainty in $R_m$ and $COV_R$ will be larger, and the confidence intervals will be larger (see Table 6). As a result, the confidence interval for the probability of failure will also be larger. We could have a larger probability of failure. The question is by how much the scope could be reduced for a given reduction in acceptable performance (i.e. for a given increase of acceptable probability of failure).

Assuming that the distribution of demand has a mean of $S_m$ and a $COV_S$ of 0.10, then, the probability of failure will depend on the mean and COV of the capacity, $R_m$ and $COV_R$. For illustration purposes only, taking a mean global safety coefficient of 3 ($R_m/S_m = 3$) and $COV_R$ equal to the confidence limit for 20 observations in the table above ($COV_R = 0.211$), a probability of failure of about 0.2% is calculated.

Reproduced from the seismic design performance targets [22], the acceptable probability of failure of an intermediate radiological hazard category nuclear installation is four times the acceptable probability for a high radiological hazard category nuclear installation, such as a traditional NPP (Table 7). For a low radiological hazard category nuclear installation, it is ten times the acceptable probability of failure in a traditional NPP.

In the previous example, if the probability of failure is computed using the $COV_R$ equal to the confidence limit for ten observations in the table above ($COV_R = 0.260$), a probability of failure of about 0.9% is obtained, that is, 4.5 times larger. Note that, in the table, the sample COV has been assumed the same for 20 and for 10 observations.

Hence, what the example shows is that, if 10 observations instead of 20 are collected, and the resulting mean and standard deviation of the sample are used as if they derived from a sample of 20 observations, then the failure probability may be about 4.5 times larger. This factor is similar to that which would be acceptable as a ratio of performance between a high and an intermediate radiological hazard installation.

TABLE 7. DESIGN PERFORMANCE TARGETS AS A FUNCTION OF RADIOLOGICAL HAZARD

| ANS 2.26 Seismic Design Categories (SDC) [23] | SSG-67 [13] | Relative probability of failure |
|---|---|---|
| SDC-5 | High radiological hazard | 0.1 |
| SDC-4 | Medium radiological hazard | 0.4 |
| SDC-3 | Low radiological hazard | 1.0 |
| SDC-2 | Conventional | 4.0 |
| SDC-1 | Conventional | 10.0 |

If the exercise is repeated with six observations, a probability of failure of about 3.5% is computed, which is 17.5 times larger than 0.2%. This ratio would only be acceptable for a conventional installation.

### I.4.4. Summary

Potential application of a graded approach to the geotechnical site characterization campaign, with respect to what would have been done for a traditional NPP is very site dependent. The exact approach will need to be defined by an experienced team of qualified professionals.

However, if a geotechnical investigation campaign is interpreted as a sampling problem, general guidance can be drawn from sampling theory and the minimum performance targets that are acceptable for the radiological hazard category of a nuclear installation.

In the simple example used to illustrate the method, it resulted that the number of geotechnical observations necessary for an intermediate radiological hazard installation could be approximately one–half of that required for a traditional NPP.

For a conventional installation (no radiological hazard), the number of observations could be less than one–third of that required for a traditional NPP.

## I.5. DERIVATION OF GEOTECHNICAL DESIGN PARAMETERS FROM A GRADED INVESTIGATION

In the derivation of design parameters from a graded investigation, the level of additional uncertainty introduced by the graded approach will need to be assessed. Note that, given site specificity, the reference uncertainty that which would have resulted, had the same rules as for a traditional NPP been applied during the site investigation.

When the increased uncertainty is judged to lead to acceptable performance for the radiological hazard category of the installation[8], then the geotechnical design parameters are determined from the site characterization in the same way as for a traditional NPP. The example in the previous section shows a simple way to assess relative performance.

[8] For an intermediate radiological hazard installation, a probability of failure less than four times the probability of failure derived from the level of uncertainty corresponding to not applying a graded approach. For a low radiological hazard installation, a probability of failure less than ten times the probability of failure derived from the level of uncertainty corresponding to no graded approach.

On the contrary, if the increased level of uncertainty leads to a performance considered unacceptable, then geotechnical design parameters determined in the same way as for a traditional NPP will need to be corrected towards the conservative side, until performance (i.e. relative probability of failure) becomes acceptable for the radiological hazard category of the installation[9].

[9] Reference [51] provides a methodology for developing these corrections based on sampling theory.

REFERENCES


[1] EUROPEAN ATOMIC ENERGY COMMUNITY, FOOD AND AGRICULTURE ORGANIZATION OF THE UNITED NATIONS, INTERNATIONAL ATOMIC ENERGY AGENCY, INTERNATIONAL LABOUR ORGANIZATION, INTERNATIONAL MARITIME ORGANIZATION, OECD NUCLEAR ENERGY AGENCY, PAN AMERICAN HEALTH ORGANIZATION, UNITED NATIONS ENVIRONMENT PROGRAMME, WORLD HEALTH ORGANIZATION, Fundamental Safety Principles, IAEA Safety Fundamentals No. SF-1, IAEA, Vienna (2006), https://doi.org/10.61092/iaea.hmxn-vw0a.

[2] INTERNATIONAL ATOMIC ENERGY AGENCY, "Site Evaluation for Nuclear Installations," IAEA Safety Standards Series No. SSR-1, IAEA, Vienna, (2019).

[3] INTERNATIONAL ATOMIC ENERGY AGENCY, Site Survey and Site Selection for Nuclear Installations, IAEA Safety Standards Series No. SSG-35, IAEA, Vienna (2015).

[4] INTERNATIONAL ATOMIC ENERGY AGENCY, Hazards Associated with Human Induced External Events in Site Evaluation for Nuclear Installations, IAEA Safety Standards Series No. SSG-79, IAEA, Vienna (2023).

[5] INTERNATIONAL ATOMIC ENERGY AGENCY, Meteorological and Hydrological Hazards in Site Evaluation for Nuclear Installations, IAEA Safety Standards Series No. SSG-18, IAEA, Vienna (2011).

[6] INTERNATIONAL ATOMIC ENERGY AGENCY, Seismic Hazards in Site Evaluation for Nuclear Installations, IAEA Safety Standards Series No. SSG-9 (Rev. 1), IAEA, Vienna (2022).

[7] INTERNATIONAL ATOMIC ENERGY AGENCY, Geotechnical Aspects in Siting and Design of Nuclear Installations, SSG-93, revision to IAEA Safety Standards Series No. NS-G-3.6, IAEA, Vienna (in preparation).

[8] INTERNATIONAL ATOMIC ENERGY AGENCY, Investigation of Site Characteristics and Evaluation of Radiation Risks to the Public and the Environment in Site Evaluation for Nuclear Installations, SSG-92, revision to IAEA Safety Standards Series No. NS-G-3.2, IAEA, Vienna (in preparation).

[9] INTERNATIONAL ATOMIC ENERGY AGENCY, Volcanic Hazards in Site Evaluation for Nuclear Installations, IAEA Safety Standards Series No. SSG-21, IAEA, Vienna (2012).

[10] INTERNATIONAL ATOMIC ENERGY AGENCY, Guidelines for Generic Roadmap Project: Nuclear Safety Infrastructure for Licensing of First Nuclear Reactor, IAEA Safety Report Series, IAEA, Vienna (in preparation).

[11] INTERNATIONAL ATOMIC ENERGY AGENCY, Establishing the Safety Infrastructure for a Nuclear Power Programme, IAEA Safety Standards Series No. SSG-16 (Rev. 1), IAEA, Vienna (2020).

[12] INTERNATIONAL ATOMIC ENERGY AGENCY, Managing Siting Activities for Nuclear Power Plants, IAEA Nuclear Energy Series No. NG-T-3.7 (Rev. 1), IAEA, Vienna (in preparation).

[13] INTERNATIONAL ATOMIC ENERGY AGENCY, Seismic Design for Nuclear Installations, IAEA Safety Standards Series No. SSG-67, IAEA, Vienna (2021).

[14] INTERNATIONAL ATOMIC ENERGY AGENCY, Design of Nuclear Installations Against External Events Excluding Earthquakes, IAEA Safety Standards Series No. SSG-68, IAEA, Vienna (2021).

[15] NUCLEAR ENERGY INSTITUTE, Industry Guideline for Developing a Plant Parameter Envelope in support of an Early Site Permit, Rep. NEI 10-01 (Revision 1), Washington, DC (2012).

[16] IDAHO NATIONAL LABORATORY, Site Suitability and Hazard Assessment Guide for Small Modular Reactors, Rep. INL/EXT-13-29749, Idaho Natl Lab., ID (2013).

[17] INTERNATIONAL ATOMIC ENERGY AGENCY, Advanced Reactor Information System (ARIS), database of advanced reactor design descriptions, https://aris.iaea.org/.

[18] COLEMAN, J., INL Seismic 10-year Re-evaluation Process for Existing Nuclear Facilities - Following a Risk-Informed Methodology, in SMiRT 23, Special Session on Risk-Informed Decision Making, Manchester, UK (2015).

[19] INTERNATIONAL ATOMIC ENERGY AGENCY, Application of a Graded Approach in Regulating Nuclear Installations, IAEA-TECDOC-1980, IAEA, Vienna (2021).

[20] INTERNATIONAL ATOMIC ENERGY AGENCY, Terminology Used in Nuclear Safety, Nuclear Security, Radiation Protection and Emergency Preparedness and Response, IAEA Safety and Security Glossary, IAEA, Vienna (2022).

[21] INTERNATIONAL NUCLEAR SAFETY ADVISORY GROUP, Basic Safety Principles for Nuclear Power Plants / 75-INSAG-3 Rev. 1, IAEA International Nuclear Safety Advisory Group, report INSAG-12, IAEA, Vienna (1999).

[22] AMERICAN SOCIETY OF CIVIL ENGINEERS, Seismic Design Criteria for Structures, Systems, and Components in Nuclear Facilities, ASCE 43-19, Reston, VA (2019).

[23] AMERICAN NUCLEAR SOCIETY, Categorization of Nuclear Facility Structures, Systems and Components for Seismic Design, ANSI/ANS 2.26-2004, La Grange Park, IL (2017).

[24] KENNEDY, R.P., Performance-goal based (risk informed) approach for establishing the SSE site specific response spectrum for future nuclear power plants, Nuclear Engineering and Design, 2413, (2011) 648-656.

[25] UNITED STATES DEPARTMENT OF ENERGY, Hazard Categorization of DoE Nuclear Facilities, DoE Standard DOE-STD-1027-2018, USDOE, Washington, DC (2018).

[26] UNITED STATES DEPARTMENT OF ENERGY, Natural Phenomena Hazards Design and Evaluation Criteria for Department of Energy Facilities, DoE Standard DOE-STD-1020-2016, USDOE, Washington, DC (2016).

[27] UNITED STATES DEPARTMENT OF ENERGY, Natural Phenomena Hazards Analysis and Design Handbook for DoE Facilities, DOE Handbook DOE-HDBK-1220-2017, USDOE, Washington, DC (2017).

[28] NUCLEAR ENERGY INSTITUTE, Small Modular Reactor Source Terms, Position Paper, USNRC ADAMS accession number ML13004A390, NRC, Washington, DC (2012).

[29] NUCLEAR REGULATORY COMMISSION, RADTRAD: A Simplified Model for Radionuclide Transport and Removal And Dose Estimation, Rep. NUREG/CR-6604 (see also Supplements 1 and 2), Sandia Natl Lab., Albuquerque, NM, (1998).

[30] NUCLEAR REGULATORY COMMISSION, XOQDOQ: Computer Program for the Meteorological Evaluation of Routine Effluent Releases at Nuclear Power Stations, Rep. NUREG/CR-2919 (Program included in the NRC-Dose suite of programs, 2012), Pacific Northwest Lab., Richland, WA (1982).

[31] INTERNATIONAL ATOMIC ENERGY AGENCY, Format and Content of the Safety Analysis Report for Nuclear Power Plants, IAEA Safety Standards Series No. SSG-61, IAEA, Vienna (2021).

[32] NUCLEAR REGULATORY COMMISSION, Alternative Radiological Source Terms for Evaluating Design Basis Accidents at Nuclear Power Reactors, Regulatory Guide 1.183, NRC, Washington, DC (2000).

[33] AMERICAN NUCLEAR SOCIETY, Criteria for Investigations of Nuclear Facility Sites for Seismic Hazard Assessments, ANSI/ANS-2.27-2020, La Grange Park, IL (2020).

[34] AMERICAN NUCLEAR SOCIETY, Probabilistic Seismic Hazard Analysis, ANSI/ANS-2.29-2020, La Grange Park, IL (2020).

[35] UNITED STATES DEPARTMENT OF ENERGY, Guidelines for Use of Probabilistic Seismic Hazard Curves at Department of Energy Sites for Department of Energy Facilities, DoE Standard DOE-STD-1024-92, USDOE, Washington, DC (1996).

[36] BENJAMIN, J.R. and CORNELL, C.A, Probability, Statistics, and Decision for Civil Engineers, McGraw-Hill, New York, NY (1970).

[37] ELECTRIC POWER RESEARCH INSTITUTE, Methodology for Developing Seismic Fragilities, Rep. EPRI TR-103959, Palo Alto, CA (1994).

[38] KENNEDY, R.P. and SHORT, S.A., Basis for Seismic Provisions of DOE-STD-1020, Rep. UCRL-CR-111478, Lawrence Livermore Natl Lab., Livermore, CA (1994).

[39] INTERNATIONAL ATOMIC ENERGY AGENCY, Optimization of Safety Measures for Protection of Nuclear Installations against External Hazards, IAEA-TECDOC-2042, IAEA, Vienna (2024).

[40] INTERNATIONAL ATOMIC ENERGY AGENCY, Assessment of Vulnerabilities of Operating Nuclear Power Plants to Extreme External Events, IAEA-TECDOC-1834, IAEA, Vienna (2017).

[41] CHOKSHI, N., BUDNITZ, R., RAVINDRA, M.K., DASGUPTA, B. and STAMATAKOS, J., A Proposed Alternative Risk-Informed and Performance-Based Regulatory Framework for Seismic Safety at NRC Regulated Facilities - Task 3, Center for Nuclear Waste Regulatory Analyses - SWRI, San Antonio, TX (2020).

[42] INTERNATIONAL ATOMIC ENERGY AGENCY, Leadership and Management for Safety, IAEA Safety Standards Series No. GSR Part 2, IAEA, Vienna (2016).

[43] NUCLEAR REGULATORY COMMISSION, Updated Implementation Guidelines for SSHAC Hazard Studies, Rep. NUREG-2213, NRC, Washington, DC (2018).

[44] DANCIU, L., et al., The 2020 update of the European Seismic Hazard Model-ESHM20: Model Overview, EFEHR Technical Report 001, (2021).

[45] PETERSEN, M.D. et al., The 2018 update of the US National Seismic Hazard Model: Overview of model and implications, Earthquake Spectra, **36** 1 (2020) 5-41.

[46] BOMMER, J.J., and ABRAHAMSON, N.A., Why do modern probabilistic seismic-hazard analyses often lead to increased hazard estimates?, Bulletin of the Seismological Society of America, **96** 6, (2006) 1967-1977.

[47] NUCLEAR REGULATORY COMMISSION, Implications of Updated Probabilistic Seismic Hazard Estimates in Central and Eastern United States on Existing Plants – Safety/Risk Assessment, Generic Issue 199 (GI-199), NRC, Washington, DC (2010).

[48] VALENTINI, A., and BELTRAN, F., Mean-to-median spectral acceleration ratios in Senior Seismic Hazard Analysis Committee (SSHAC) Level 3 probabilistic seismic hazard analyses: An analysis of reported results, Earthquake Spectra, 40 1, (2024) 469-485.

[49] SCHNEIDER, J., Introduction to Safety and Reliability of Structures, International Association For Bridge And Structural Engineering, Rep. SED-05, Zürich (2006).

[50] CASCIATI, F., "Safety index, stochastic finite elements and expert systems", Reliability Problems: General Principles and Applications in Mechanics of Solids and Structures, International Centre for Mechanical Sciences (CISM) Courses and Lectures No. 317, Udine, Italy (1991).

[51] RUSSIAN FEDERATION FEDERAL AGENCY FOR TECHNICAL REGULATION AND METROLOGY, Soils – Methods for Statistical Processing of Test Results, GOST 20522-2012, Moscow (2012).

## Annex I

## TABLES TO BE USED IN SITING PROCESS

Table I–1 [I–1] presents an overview of the criteria typically associated with various aspects of the siting process.

In Table I–2 relevant IAEA safety standards [I–2 to I–10] are cross referenced as being applicable to the siting issues addressed in IAEA Safety Standards Series No. SSG-35, Site Survey and Site Selection for Nuclear Installations [I–1]. The requirements and recommendations within these safety standards are valuable for evaluating candidate sites.

TABLE I–1. SCREENING AND RANKING CRITERIA FOR SITE SELECTION
*(reproduced from Ref. [I–1])*

| Criteria | | Category | | |
|---|---|---|---|---|
| Primary | Type | Screening | | Ranking |
| | | Exclusionary | Discretionary | |
| Earthquake | Ground vibration | | ✓ | ✓ |
| | Surface rupture | ✓ | | |
| Geotechnical | Slope instability (massive | ✓ | | |
| | Slope instability (minor) | | ✓ | ✓ |
| | Subsidence | | ✓ | ✓ |
| | Massive liquefaction | ✓ | | |
| | Liquefaction | | ✓ | ✓ |
| | Karst (massive) | ✓ | | |
| Volcanism | Lava flow | ✓ | | |
| | Pyroclastic flow | ✓ | | |
| | Ground deformation | ✓ | | |
| | Tephra fall | | ✓ | ✓ |
| | Volcanic gases | | ✓ | ✓ |
| | Lahars (massive) | ✓ | | |
| Flooding | River | | ✓ | ✓ |
| | Dam break | | ✓ | ✓ |
| | Coastal (storm surges, waves, | | ✓ | ✓ |
| | Tsunami | | ✓ | ✓ |

TABLE I–1. SCREENING AND RANKING CRITERIA FOR SITE SELECTION
*(reproduced from Ref. [I–1]) (cont.)*

| Criteria | | Category | | |
|---|---|---|---|---|
| Primary | Type | Screening | | Ranking |
| | | Exclusionary | Discretionary | |
| Extreme meteorological events | High straight winds | | ✓ | ✓ |
| | Tornadoes | | ✓ | ✓ |
| | Tropical storms | | ✓ | ✓ |
| | Precipitation | | ✓ | ✓ |
| | Sand storms and dust storms | | ✓ | ✓ |
| Human induced events | Aircraft crash | | ✓ | ✓ |
| | Explosions | | ✓ | ✓ |
| | Gas releases | | ✓ | ✓ |
| | External fires | | ✓ | ✓ |
| | Electromagnetic interference | | ✓ | ✓ |
| Nuclear security events | | | ✓ | ✓ |
| Dispersion | In air and water | | ✓ | ✓ |
| Feasibility of implementation of emergency plan | | ✓ | | |
| Implementation of emergency plan | | | ✓ | ✓ |
| Non-safety | Topography | | ✓ | ✓ |
| | Availability of cooling water | ✓ | ✓ | ✓ |
| | Access to water | | ✓ | ✓ |
| | Availability of transport | | ✓ | ✓ |
| | Access to national or regional electricity grid | | ✓ | ✓ |
| | Non-radiological environmental impacts | ✓ | ✓ | ✓ |
| | Socioeconomic impacts | | ✓ | ✓ |
| | Land use planning | | ✓ | ✓ |

TABLE I–2. SITE SELECTION ISSUES CROSS REFERENCE TO IAEA SAFETY STANDARDS
*(reproduced from Ref. [I–1])*

| Site selection issues | | Safety Requirements | Site evaluation Safety Guides | | | | | | Design Safety Guides | |
|---|---|---|---|---|---|---|---|---|---|---|
| Primary | Effect | SSR-1 | SSG-79 | SSG-92 | SSG-9 (Rev. 1) | SSG-18 | SSG-21 | SSG-93 | SSG-68 | SSG-67 |
| Earthquake | Ground vibration | ✓ | | | ✓ | | | | | ✓ |
| | Surface rupture | ✓ | | | ✓ | | | | | |
| Geotechnical | Slope instability | ✓ | | | | | | ✓ | | |
| | Subsidence | ✓ | | | | | | ✓ | | |
| | Soil liquefaction | ✓ | | | | | | ✓ | | |
| | Extensive oil and gas extraction history | ✓ | | | | | | ✓ | | |
| Volcanism | | ✓ | | | | | ✓ | | | |
| Flooding | River | ✓ | | | | ✓ | | | ✓ | |
| | Dam break | ✓ | | | | ✓ | | | ✓ | |
| | Coastal | ✓ | | | | ✓ | | | ✓ | |
| | Tsunami | ✓ | | | | ✓ | | | ✓ | |
| Extreme meteorological events | High straight winds | ✓ | | | | ✓ | | | | |
| | Tornados | ✓ | | | | ✓ | | | ✓ | |
| | Precipitation | ✓ | | | | ✓ | | | ✓ | |
| Human induced events | Aircraft crash | ✓ | ✓ | | | | | | ✓ | |
| | Explosions | ✓ | ✓ | | | | | | ✓ | |
| | Gas releases | ✓ | ✓ | | | | | | ✓ | |

TABLE I–2. SITE SELECTION ISSUES CROSS REFERENCE TO IAEA SAFETY STANDARDS
*(reproduced from Ref. [I–1]) (cont.)*

| Site selection issues | | Safety requirements | Site evaluation Safety Guides | | | | | | Design Safety Guides | |
|---|---|---|---|---|---|---|---|---|---|---|
| Primary | Effect | SSR-1 | SSG-79 | SSG-92 | SSG-9 (Rev. 1) | SSG-18 | SSG-21 | SSG-93 | SSG-68 | SSG-67 |
| | External Fires | ✓ | ✓ | | | | | | ✓ | |
| Population | Density | ✓ | | ✓ | | | | | | |
| | Distance from centres | ✓ | | ✓ | | | | | | |
| Dispersion | In air | ✓ | | ✓ | | | | | | |
| | In water | ✓ | | ✓ | | | | | | |
| Feasibility of the emergency plan | | ✓ | | ✓ | | | | | | |

## References to Annex I


[I–1] INTERNATIONAL ATOMIC ENERGY AGENCY, Site Survey and Site Selection for Nuclear Installations, IAEA Safety Standards Series No. SSG-35, IAEA, Vienna (2015).

[I–2] INTERNATIONAL ATOMIC ENERGY AGENCY, Site Evaluation for Nuclear Installations, IAEA Safety Standards Series No. SSR-1, IAEA, Vienna (2019).

[I–3] INTERNATIONAL ATOMIC ENERGY AGENCY, Hazards Associated with Human Induced External Events, IAEA Safety Standards Series No. SSG-79, IAEA, Vienna (2023).

[I–4] INTERNATIONAL ATOMIC ENERGY AGENCY, Investigation of Site Characteristics and Evaluation of Radiation Risks to the Public and the Environment in Site Evaluation for Nuclear Installations, SSG-92, revision to IAEA Safety Standards Series No. NS-G-3.2, IAEA, Vienna, (in preparation).

[I–5] INTERNATIONAL ATOMIC ENERGY AGENCY, Seismic Hazards in Site Evaluation for Nuclear Installations, IAEA Safety Standards Series No. SSG-9 (Rev. 1), IAEA, Vienna (2022).

[I–6] INTERNATIONAL ATOMIC ENERGY AGENCY, WORLD METEOROLOGICAL ORGANIZATION, Meteorological and Hydrological Hazards in Site Evaluation for Nuclear Installations, IAEA Safety Standards Series No. SSG-18, IAEA, Vienna (2011) (revision in preparation).

[I–7] INTERNATIONAL ATOMIC ENERGY AGENCY, Volcanic Hazards in Site Evaluation for Nuclear Installations, IAEA Safety Standards Series No. SSG-21, IAEA, Vienna (2012).

[I–8] INTERNATIONAL ATOMIC ENERGY AGENCY, Geotechnical Aspects in Siting and Design of Nuclear Installations, SSG-93, revision to IAEA Safety Standards Series No. NS-G-3.6, IAEA, Vienna (in preparation).

[I–9] INTERNATIONAL ATOMIC ENERGY AGENCY, Design of Nuclear Installations Against External Events Excluding Earthquakes, IAEA Safety Standards Series No. SSG-68, IAEA, Vienna (2021).

[I–10] INTERNATIONAL ATOMIC ENERGY AGENCY, Seismic Design for Nuclear Installations, IAEA Safety Standards Series SSG-67, IAEA, Vienna (2021).

## Annex II

## DATA NECESSARY FOR SITE SURVEY AND SITE EVALUATION

### II–1. TOPICS PRIMARILY RELATED TO SAFETY AND HEALTH ASPECTS

#### II–1.1. Geology and hydrogeology

The geological environment includes the bedrock and overburden geology, notably the aquifers, at both the local and regional scales. Paragraph 6.3 of IAEA Safety Standards Series No. SSG-92, Investigation of Site Characteristics and Evaluation of Radiation Risks to the Public and the Environment in Site Evaluation for Nuclear Installations (a forthcoming revision to [II–1]) states:

> "The hydrosphere is a major medium by which radioactive material that is released from a nuclear installation via discharges or accidental releases could be dispersed into the environment and transferred to locations where water is used by or for the population."

Thus, sites with a simple structural context and hydrogeological functioning need to be sought so that their conceptual and numerical representation, necessary for the analysis of a possible migration of radionuclides, is made possible and reliable.

The geological characteristics of the site play a major role in site evaluation because they form a basic element in the conditions for radionuclide migration and in several natural external hazards (e.g. geotechnical hazard, seismic, flooding, volcanology). The basic data needed for siting gathers the geological units, the stratigraphy, their three dimensional geometry and extent (e.g. through maps, cross-sections), their mineralogy and petrology, their hydraulic, geomechanical and geotechnical properties. Such data may be collected through fieldwork, boreholes and geophysical surveys (e.g. see [II-2 to II-4]). The structural geology as well as a description of the fractures and faults and their characteristics are necessary both to account for potential direct bypasses in the modelling of radionuclide migration and for evaluation of seismic hazard (see Section II-1.2 below). The potential for surface faulting or deformation may also concern the design of the facility and structures, systems and components, as the design has to withstand the effects of any site or near-site surface faulting or deformation, which can be due to tectonic or non-tectonic structures. The retention properties of the geological units and the characteristics (e.g. geometry, hydraulic parameters, geochemistry) [II–2] of the groundwater bodies, including data on the connectivity between aquifers, are also primary data to simulate the flow of radionuclides in normal situations as well as in situations resulting from accidents (e.g. [II–4]).

The hydrological regimes, seasonal, yearly and long term variations of the main aquifers from available piezometric measurements are necessary, notably for the evaluation of the external flooding hazard (for recommendations on how to assess the flooding hazard see IAEA Safety Standards Series No. SSG-18, Meteorological and Hydrological Hazards in Site Evaluation for Nuclear Installations [II–5]). In particular, para. 3.31 of [II–5] states:

> "Long term records of groundwater levels should be obtained from data on water levels in wells in the same region and in comparable hydrogeological situations to allow estimation of the effects of extreme meteorological conditions on groundwater levels (to observe the effects of storm events) and to examine long term trends such as trends due to large scale groundwater extraction."

Additional geology and hydrogeology data other than the above mentioned elements are needed to provide a comprehensive description of the site's initial state and its evolution due to construction and operation of a nuclear installation. This includes information on the existing ambient water quality in the study areas, with a focus on characterizing both radiological and non-radiological species (see [II–4, II–6]). In addition, any changes to the geology and geomorphology resulting from the nuclear installation that may result in a change of natural hazards have to be described, such as removal of bedrock and/or unconsolidated deposits, disturbance of soils or sediments such as stockpile, changes made that would affect coastal processes and features or any modification in the interrelationships with the groundwater regime (e.g. [II–4]).

Information on hydrogeological conditions is also relevant to the assessment the dispersion of radionuclides through the environment associated with normal operation and accident conditions at the nuclear installation, as well as associated with the transport of radioactive material to and from the site, if any (see [II–7]).

The description of the human current use of natural resources (mineral, geothermal, groundwater) in the site area and of potential uses for the future is necessary for evaluating the impact of radioactive and non-radioactive releases to the public (due to ingestion and inhalation) resulting from normal operation and from accident conditions (see [II–5]). Anthropogenic influences on the groundwater levels, such groundwater extraction, artificial recharge and backfill are also necessary for the modelling of the radionuclide migration to the environment (see [II–5]). These data on the human use of natural resources are also needed to adapt the general design (including depth and location of the nuclear installation in the site) and establish measures for site protection (dams, drainage network and extent of the buffer zone). As an example, the use of natural resources in the region around the nuclear installation site might not be compatible with the siting of the installation without defining a strict buffer zone where such activities are forbidden.

This geological and hydrogeological database can be helpful in checking the site selection criteria, such as safety-related criteria associated with the dispersion in groundwaters.

Finally, the hydrogeological characteristics of the site are needed to develop and implement a relevant monitoring plan, both allowing to detect any abnormal groundwater level prone to generate flooding and to improve the models of dispersion of radionuclides (see [II–1]). The knowledge of the site hydrogeology and associated monitoring device are also an essential part of the development of the emergency planning.

### II–1.2. Regional and local tectonics, including fault displacement and seismology

Tectonics is the study of the structure of the Earth's crust and its deformations. It is a key process in the understanding of several geological phenomena that might affect the safety of nuclear installations, such as vertical movements (erosion, landslide hazard) and the opening of natural fractures (and thus their hydraulic properties). But most of all, the tectonic activity may give rise to earthquakes and associated soil shaking and potentially, fault displacement. More precisely, seismology is the study of earthquakes and the propagation of elastic waves through the Earth.

To evaluate the seismic hazard at the site and surrounding area, it may be necessary, in addition to the historical seismic data (subdivided into pre-instrumental and instrumental temporal scale) to collect data on the potential prehistorical seismic events. Paragraph 3.24 of SSG-9 (Rev. 1) [II–2] provides guidance on investigating regional features to help characterize prehistoric seismicity. A graded approach can be applied to the number and variety of data needed for evaluating the seismic hazard to the targeted nuclear installation type; an example is given in Section 4. Paragraph 3.17 of SSG-9 (Rev. 1) [II–2] states that "The size of the region to be investigated when assessing vibratory ground motion hazards should be large enough to incorporate all seismogenic structures that could affect the nuclear installation."

Selecting a site in a stable geological environment is beneficial, so that site effects due to natural phenomena that may occur (vertical movements, seismicity, neotectonics) remain acceptable for installation safety. For this, a site distant from tectonic plate boundaries and major faults is preferred.

A geological map of the site, obtained by geological mapping, borehole data and geophysical surveys (see Section II–1.1), is also necessary to locate large faults, either outcropping or hidden below ground. Requirement 15 of IAEA Safety Standards Series No. SSR-1, Site Evaluation for Nuclear Installations [II–8] states that "**Geological faults larger than a certain size and within a certain distance of the site and that are significant to safety shall be evaluated to identify whether these faults are to be considered capable faults.**" Additional data on the tectonic history of the region or the present stress field may also give information on the properties (hydraulic, mechanical) of these large faults.

These geological data should be used to identify the potential seismic events of relevance for the plant's safety in the site area, including seismogenic probability and recurrence rates. The list may include the event of an earthquake occurring on an undetected (hidden) fault below the site. IAEA Safety Standards Series No. SSG-67, Seismic Design for Nuclear Installations [II–9] provides recommendations on determining the design basis ground motion (safe shutdown earthquake) for the site based on the possible seismic sources. The seismicity potential due to human induced activities (explosions or vibrations due to transport, industrial and other installations in the site vicinity) also needs to be considered.

Evaluation of seismic effects on the plant is also needed so as a design input. The possible challenges posed to the nuclear installation safety need to be evaluated in terms of ground motion, fault displacement hazards or any modification of groundwater flows. SSG-67 [II–9] provides recommendations on how to adapt the design of nuclear installations subjected to seismic hazards, evaluated in accordance with SSG-9 (Rev. 1) [II–2]. It specifically provides recommendations on the necessary data obtained from the geological, geophysical and geotechnical investigations, including static and dynamic soil properties as well as the variability in the thickness and properties of subsurface material layers.

The seismic hazard and the existence of capable faults contribute to the site selection criteria related to natural hazards. Also, the potential consequences relating to dispersion in air and water (including both surface and groundwater) in the case of earthquakes may be associated with non-safety related site selection criteria (see [II–10]). Otherwise, the plant is to be designed to protect against direct (shaking, fault displacement) and indirect (dam failure, tsunami) effects of earthquakes (basically, siting away from faults, earthquake-resistant works, safe shutdown of the plant (see [II–9])).

Early in the project, seismic monitoring instruments should be placed at and around site to detect and record ground motions and develop the local seismicity database. These instruments should be maintained throughout the nuclear installation's lifetime to provide the information necessary for operator actions in the case of seismic events. The seismotectonic data for the site should be updated throughout the nuclear installation's lifetime, notably on the basis of monitoring.

Finally, data should be provided to assess the feasibility of developing an emergency response plan for seismic events at the nuclear installation site.

### II–1.3. Volcanology

Paragraph 1.16 of IAEA Safety Standards Series No. SSG-21, Volcanic Hazards in Site Evaluation for Nuclear Installations [II–11] states that "Volcanism is the natural process resulting from magma ascending through the Earth and erupting, or nearly erupting, at the Earth's surface and producing phenomena that may have far reaching and long term effects." The possibility of such disruptive phenomena occurring in the vicinity of site needs to be considered early in the site survey and evaluation stages to determine whether suitability of the site can be confirmed. Thus, ideally, the best site would have no indication of any past or present volcanic activity. Otherwise, different types of phenomena associated with volcanism should be considered for their impact on site suitability and derivation of design basis parameters.

The hazards posed by volcanic activity that could potentially impact the safety of a nuclear installation have to be thoroughly evaluated. Given that certain volcanic phenomena can impact sites located hundreds of kilometres from an erupting volcano, it is crucial to apply a comprehensive methodology for assessing volcanic hazards. However, in many areas in the world it can be easily demonstrated on the sole basis of available geological knowledge during the initial assessment that no volcanic activity occurred in the past ten million years in a large region around the potential site: in that case, the volcanic hazard can be screened out. If this is not the case, capable volcanoes are to be identified and evaluated (see [II–11]), along with their associated hazards, to provide the necessary input for determining design basis parameters or to re-evaluate the site suitability. This information is also essential for conducting deterministic and/or probabilistic safety assessments throughout the lifetime of the nuclear installation.

Volcanic hazards are complex and varied. Paragraph 1.16 of SSG-21 [II–11] states that "the potential for various flow phenomena, such as pyroclastic flows or lava flows, within the site vicinity should be assessed as part of the site suitability evaluation." Various external conditions such as the topography of the site and surrounding region and the weather conditions during a potential volcanic event may have to be accounted for in the evaluation of the hazard. The impact of volcanic phenomena, combined with other volcanically induced hazards such as volcanic tephra fall or water impoundments caused by debris flows and lava flows leading to flooding, need to be considered. In addition, para. 1.15 of SSG-21 [II–11] states that "some hazards can even be present at inactive volcanoes, for example, the potential for collapse of a volcano edifice generating a tsunami long after volcanic activity has ceased." SSG-21 [II–11] provides recommendations on assessing volcanic hazards at a site for a nuclear installation, aimed at enabling the identification and thorough characterization of all possible hazardous phenomena that could be associated with volcanic events in future.

Paragraph 5.8 of SSR-1 [II–8] states:

> "A proposed new site shall be considered unsuitable if reliable evidence shows the existence of a capable volcano that has the potential to affect the safety of the nuclear installation and which cannot be compensated for by means of a combination of measures for site protection and design features of the nuclear installation."

Paragraph 1.16 of SSG-21 [II–11] states that "the opening of a new volcanic vent, may create a hazard (or hazards) that represents an exclusion condition and, therefore, precludes the particular nuclear installation site from further consideration during site selection." The potential for other volcanic phenomena, such as the accumulation of volcanic tephra, may need to be considered as design basis external events. Thus, a site specific volcanological database should be developed with reference data and associated uncertainties, notably in age determinations, on geological and historic volcanic activities on a large region around the site (i.e. 100 km radius). Consequently, any present data on indirect indices or consequences of present or past volcanic activities in the site vicinity need to be collected: gas releases, high geothermal gradient, former volcano edifice causing instabilities (collapse), modifications in surface and groundwater chemical properties.

Engineering solutions or operational measures can generally be employed to mitigate some of the external event's effects by incorporating specific design features and/or implementing operational procedures. Measures for protecting the whole site may be established, as well as monitoring of key parameters (to be defined) that would allow detection of any volcanic activity having potential consequences on safety and the development of an emergency response plan. However, if these solutions are not feasible or cannot be proven sufficient to mitigate effects of external events, or if the feasibility of developing an emergency response plan for volcanic events at the nuclear installation site cannot be assessed, an alternative site needs to be selected.

### II–1.4. Subsoil and geotechnical conditions

Geotechnics is the study of soils (in the mechanical sense of the term, as opposed to rock) in direct relation to the construction of the plant's works. It embraces the applications of soil mechanics, rock mechanics and engineering geology. The geotechnical investigation programme covers, in addition to general information on geology and groundwater (see Section II–1.1), the description of the extent and nature of subsurface materials as well as the soil and rock properties (see [II–3]), to provide necessary information or primary data for making decisions regarding the suitability and characteristics of subsurface materials.

The site geotechnical studies essentially aim at evaluating and reducing the potential for soil instabilities and geotechnical hazards like landslide, collapse, soil swelling, soil liquefaction, subsidence or uplift, collapse of cavities, in the site and surrounding area (e.g. [II–12]). Such evaluation may involve collecting data on the potential past geotechnical events. The sources of data are initially historical and current documents, completed as needed by a new geotechnical investigation programme comprising in situ exploration and laboratory tests (see [II–3]). This programme notably includes a study of the foundation material's stability under static, seismic and dynamic loadings. A soil and rock profile needs to be developed for the site, considering the variability and uncertainty of the subsurface materials. The geological map of the site, obtained by fieldwork, shallow surface borehole data and/or geophysical surveys (see Section II–1.1), provides information on the nature of the soils and presence of natural features that might influence both the surface and subsurface stability of the nuclear installation site. Information on groundwater (regime, location, characteristics, physical chemistry) is also needed.

These data will be used to compile a list of potential events that could impact the safety of plant due to site instability within the site area. This list will include the credibility and recurrence probability of these events, their ability to affect simultaneously multiple units at the site, and their possible effects on the nuclear installation based on site specific data. This may include the stability of slopes under static, dynamic and seismic loading, being natural or human-made (mines, cut slopes, dams and embankments, waste stockpile), magnitude and rate of foundations and/or soils settlement due to loads transferred from structures and/or changes in groundwater level, differential settlement and soil distortion, subsurface instability (collapse, subsidence or uplift) and groundwater inflow. SSG-93 [II–3] provides recommendations on the evaluation of geotechnical hazards. The potential for geotechnical hazards due to human activities (modifications in natural network, past mining activities) needs also to be considered.

The geotechnical characteristics of the site are also necessary for evaluation of other hazards such as flooding and earthquakes (see [II–10]).

An evaluation of the effect of identified potential geotechnical hazards on the plant is also needed. The potential impacts on the safety of the nuclear installation need to be assessed. In certain cases, the complex subsurface conditions can be treated, for example by compacting the soil or filling the cavities.

The geotechnical hazard would contribute to the site selection criteria related to natural hazards, by rejection of the sites having unacceptable subsurface conditions or by ranking the sites with regard to their subsurface conditions. Also, the potential consequences in relation to dispersion in air and water (both surface groundwater) in case of a geotechnical event may be associated with non-safety related site selection criteria. Otherwise, the design of the installation must protect against the effects of geotechnical events.

Monitoring devices need to be installed at the site to gather data on various parameters and site characteristics, particularly to detect any changes in the soil characteristics, such as sliding or vertical movement. These devices need to be in placed early in the project and continue to operate throughout the lifetime of the nuclear installation (see [II–3]), providing necessary information for operator actions in response to geotechnical events. It could be completed with monitoring of key parameters (to be defined) that would allow detecting any unexpected evolution having potential consequences on safety. The knowledge of the geotechnical characteristics of the site needs to be updated throughout the nuclear installation's lifetime, notably on the basis of monitoring.

Finally, data need to be provided to assess the feasibility of developing an emergency response plan for geotechnical events at the nuclear installation site.

### II–1.5. Surface water environment and hydrology

The surface water environment includes all surface water features and hydrology that might affect surface water at the site or in the local and regional study areas. Water bodies can include freshwater streams, lakes and impoundments, wetlands, sea in case of coastal sites, [II–4]. Hydrology is the science that studies occurrence, movement, distribution and properties of the Earth's water, as well as its interaction with the environment throughout the various phases of the hydrological cycle (evaporation and transport into the atmosphere and back to the land and oceans and transport through the aquatic environment).

The baseline information on the surface water environment needs to be sufficient to support the use of an aquatic dispersion model to conduct the site specific safety analysis and environmental impact assessment and to support the evaluation of environmental impacts on the nuclear installation, particularly in relation to flooding.

The project activities are anticipated to change the surface water environment [II–4]. In particular, the initial state of the surface water quality (e.g. water temperature, conductivity, pH, chemistry, dissolved contents and concentrations of metals) needs to be defined and assessed throughout all licensing stages. Information needs to be provided also on the natural concentrations of radionuclides and toxic chemicals in all the water bodies before the construction of the installation to identify baseline concentrations (see [II–1]). To assess the potential changes in surface water quality due to nuclear activities, all liquid effluents that could be generated during all stages in the lifetime of the facility or activity need to be identified, in particular the average and maximum emissions from point sources (concentrations/activity levels and volumes), the planned discharges and fugitive releases, the deposition from airborne particulates and the surface runoff. The sediment quality of all water bodies to be affected by the facility or activity need to be established, demonstrating the use of appropriate sampling and analytical protocols, for the analytical parameters with the potential to be influenced by the facility or activity. This information should provide an appropriate understanding of sediment characteristics and conditions at the site and within local and regional study areas. The construction of a nuclear installation can also significantly impact drainage characteristics of the site (see [II–10], i.e. surface water natural drainage routes, depth of groundwater table and the ability of surface water to flow through the site area). Also, the nuclear installation may need a water supply (for cooling during plant operation and normal shutdown, for the ultimate heat sink and for fire protection) that may exceed the water availability during low flow conditions (e.g. [II–12]). Thus, the site drainage initial characteristics both during periods of low flow and high flow need to be ascertained as well as their potential modification by the nuclear installation, for example through the development of hydrological modelling.

Conversely, the surface water environment can impact the safety of the nuclear installation by causing flooding at the site. Therefore, it is necessary to assess flood hazards, taking into account both natural and human induced events, as well as their possible combinations. External flooding may be due to earthquakes, volcanic activity, abrupt landslide or structural failure phenomena, or due to hydrometeorological phenomena such as resulting from precipitation or snowmelt, abnormal ice effects, high astronomical tide, seiches, storm surge, wind wave and tsunamis (e.g. [II–5]). Other sources of external flooding may be caused by anthropogenic activities in the surrounding area, such as a dam or coastal dyke failure. To determine the design basis flood caused by single event or a combination of such events (for details on how to assess the flooding hazard, see [II–5]), the applicant needs to collect the existing data on the variations of water levels of all relevant water bodies at the site and in surrounding areas (including their seasonal variations, historical records on minimum and maximum levels, wave characteristics, inundation events) Then, depending on the available data, it may be necessary to carry out a detailed investigations programme to characterize the hydrological hazards at both the regional and site levels. This programme needs to employ both deterministic and probabilistic methods, as appropriate, to assess normal flow, flooding and drought conditions of water bodies, as well as the interactions between surface water and the groundwater regime. This programme also needs to account for predictions of changes to surface water hydrology at the site that are anticipated due to foreseeable upstream land use changes, as well as the possible effects in the climate change due to human activities, notably indirectly by a change in meteorology (see Section II–1.6 below). SSG-18 [II–5] provides specific recommendations for assessing hydrological hazards and associated information and investigations needed.

The hydraulic characteristics of the site and surrounding region are analysed to assess the possible movement of radionuclides in surface water, which is then used to evaluate the subsequent radiological impact (see [II–1]). In the same way, the other impacts of such discharges, such as emission of non-radioactive hazardous species, changes in temperature and in the chemical properties of the natural waters, need to be evaluated. To achieve this, it is necessary to provide information on the basic physical and chemical characteristics of both natural and artificial water bodies. This includes delineating the drainage basins and compiling prehistoric, historic and instrumentally observed hydrological data, e.g. flow rates and water levels [II–4]. Additionally, it is important to investigate the hydrological regimes throughout the drainage basin, including seasonal fluctuations and annual variability of all surface waters [II–4]. A description of the main water control structures and water intake points and different uses of water in the region (drinking, agriculture, industry) will also be necessary. This information needs to facilitate the evaluation of measures taken to prevent the release of radionuclides into the environment. It is also a basis for determining the best location and the different types of instruments for monitoring the hydrosphere. SSG-92 [II–1] provides recommendations for evaluating the transport of radionuclides in the hydrosphere released from the nuclear installation and on the needed baseline.

These data are also valuable for assessing the adequacy of the surface water environment with the siting of the nuclear installation, which can be included in the site selection criteria. Information on sources of surface water and groundwater, water quantity and quality, and the reliability and availability of water supply will inform decisions about water supply needs, including potential additional requirements for emergency cooling, firefighting or other process. The evaluation also needs to consider the potential impacts of debris and fouling on the water supply for the nuclear installation [II–4].

### II–1.6. Meteorology and atmospheric environment

The atmospheric environment of the site needs to be well known, given that para. 4.2 of SSG-92 [II–1] states, "the atmosphere is a major exposure pathway by which radioactivity that is either routinely discharged under authorization or accidentally released from a nuclear installation could be transported to locations where the public can be exposed." Thus, ideally, the best site would have mild weather conditions, with low frequencies and severities of extreme meteorological hazards.

Meteorology is the study of the atmosphere and motions within the atmosphere on short timescales. Meteorology needs to be distinguished from climate, which is concerned with the long term; however, the climate, in particular the human induced climate change (see [II–5]), may have local and regional effects on the site meteorology. Several weather elements describe the atmosphere such as air temperature, relative humidity, precipitation amount and type (rainfall levels, snow loads), wind direction and speed, atmospheric pressure, solar radiation and cloud cover.

Variations in such weather elements may give rise to extreme meteorological hazards due to extreme values of these variables (temperature, precipitation) as well as to infrequent meteorological events, in particular tornadoes, hurricanes, blizzards, dust and sandstorms, hail, ice storms, lightning and drought which may impact the nuclear installation's safety. In accordance with SSR-1 [II–08], the potential for such extreme (paras 5.11, 5.12) and rare (paras 5.13, 5.14) meteorological events and their possible combinations are required to be evaluated. SSG-18 [II–5] provides detailed recommendations on the procedures for assessing hazards related to extreme meteorological events and rare meteorological phenomena. It also outlines necessary information and investigations needed for such assessments.

Meteorological hazards need to be accounted for in the physical protection design, as mentioned in Ref. [II–4]. The design requirements for physical protection are significantly influenced by the location of the site. The site evaluation should consider both the physical dimensions of the nuclear installation and the characteristics of its surrounding environment. Weather conditions that could potentially hinder the operability of physical protection systems — those systems designed to monitor reactor operations and automatically initiate preventive actions upon detecting an abnormal condition — need to be carefully considered.

To assess hazards associated with meteorological events, data need to be collected on both extreme (minimum and maximum) and average meteorological variables. This should be carried out for both the regional level and the site through an on-site meteorological monitoring station or utilizing data from existing meteorological stations. Additionally, prehistoric, historic and instrumental climate data that accurately reflect regional conditions need to be collected.

In addition to the abovementioned weather elements, comprehensive data on the atmospheric environment is needed to fully describe the initial site conditions and their evolution throughout the lifetime of the nuclear installation, including construction, operation, decommissioning and the institutional control period. This includes the background (initial) air quality in the study areas (see [II–1]), with emphasis on the characterization of radioactive and non-radioactive substances, as well as ambient daytime and nighttime noise levels at the site and local study areas. Information on its source(s), geographical extent and temporal variations need to be included. The typical natural variation in concentrations of radioactive and other hazardous substances at reference sites, along with associated uncertainties, needs to be described and documented [II–4]. Information needs to be provided on the potential emissions of hazardous and nuclear materials from the nuclear installation into the atmosphere during normal operations, as well as in postulated accident and malfunctioning scenarios, along with an assessment of resulting effects on air quality. One should consider the effect of transport associated with the nuclear installation activity (notably increased traffic along transportation corridors to and from the site during construction) and the potential impact on residences and sensitive sites such as hospitals, schools, day-care facilities, seniors' residences, and places of worship [II–4]. SSG-92 [II–1] specifically provides (in section 4) recommendations on the evaluation of the atmospheric transport of radionuclides released from the nuclear installation and on the baseline measurements needed. If wet cooling towers are planned for a nuclear installation, it may be necessary to consider dust generated by transport and operation activities [II–6].

Weather conditions are also crucial for evaluating the dispersion of radionuclides in the environment, both during normal operation and in event of an accident at the nuclear installation. This information is also relevant for assessing the dispersion related to the transport of radioactive material to and from the site, if applicable. The assessment of site meteorology, particularly wind speed and direction, is essential for developing effective emergency planning. To achieve this, monitoring devices need to be installed on site, in the surrounding region and potentially in neighbouring countries. These data are essential for guiding operator actions during meteorological events and for developing models to predict the dispersion of radionuclides (see [II–1]). Meteorological data alone are generally not sufficient to screen a site out from further consideration during site selection. Measures can often be implemented to protect safety-related equipment at the site (see [II–10]). However, specific criteria based on meteorological characteristics may be used to rank and compare candidate sites.

### II–1.7. Topography and geography

Some elementary components of the site database are the natural topography and geography and their expected alteration due to construction of the nuclear installation, such as changes in elevations and in the drainage conditions. This information is needed to assess the potential impacts of the construction, operation and closure of the nuclear installation on flora and fauna, the nuclear installation's safety itself and the sustainability of other human activities (industrial, agricultural, cultural, recreational and tourism activities) in the surrounding area. Additionally, it helps in optimizing these modifications to minimize adverse consequences.

A detailed knowledge of the site topography may be necessary for evaluating several types of natural hazards, such as the slopes for evaluating a risk of landslide, the entrenchment of rivers regarding flooding hazards, the coastal topography to evaluate some meteorological hazards, the potential for site effects in case of earthquake. Generally, a site with mostly flat ground or low altitude variations has less potential for such hazards or less consequences of such hazards and thus should be preferentially selected. The general geography of the site completes this information notably with information on land use and the vegetation cover (e.g. dip slopes in forest will be more stable than bare soil).

Such evaluation of natural hazards involves describing the general topography (including topographical maps, reference vertical datum and horizontal datum) and the general geographical features in the site vicinity, as well as the detailed topography of the areas that could be flooded, the boundaries of the watershed, the flood plain characteristics and the bathymetric data (see [II–5]). Detailed topographic features can be helpful for identifying some geological structures such as faults (see Section II-1.1). Additionally, artificial structures in the vicinity of the site, such as levees, should be described. Finally, the historical natural modifications of the topography (due to landslides, recent earthquakes, channel migrations, river captures, quick erosion or river incision) need to be investigated to assess the associated natural hazards. The regional topography can also affect the climate and meteorology of the site area. Thus, topographic knowledge is necessary for understanding the local meteorology, and thus evaluating the meteorological hazards (see Section II–1.6). This is also necessary for determining the appropriate region and specific locations to place monitoring devices for the collection of data on air and water dispersion conditions at release points.

The examination of the topography and the geography of the area surrounding the site helps to identify some potential hydrological connections between surface water and groundwater (see Sections II–1.1 and II–1.5). More generally, data on the topography and the geography are needed for developing numerical models of the radionuclide transport to the environment.

Finally, topography and geography are considered when assessing the feasibility of implementing an emergency response action plan for the nuclear installation. The feasibility, in the geographical context of the site, can be a site selection criterion for exclusion or for ranking the potential sites (see Section 2.3). As an example, a site isolated by mountains or located on a peninsula or small island can be considered as unfavourable for developing an emergency plan. Paragraph 4.9 of SSG-35 [II–10] mentions that the site topography may be one of the non-safety related criteria for site selection, especially for ranking of candidate sites, if needed.

### II–1.8. Demography

As stated in IAEA Safety Standards Series No. SF-1, Fundamental Safety Principles [II–13], "**The fundamental safety objective is to protect people and the environment from harmful effects of ionizing radiation.**" Thus, the ideal site hosting a nuclear installation with potential for radioactive releases beyond the site boundaries is to be devoid of population and human activities. However, this situation is almost non-existent and certain nuclear installations (particularly SMRs) may expressly target a service to the local population. The human presence around the site is therefore generally unavoidable and it is therefore necessary to check that they would not be adversely affected by nuclear activity and that, conversely, they would not jeopardize the safety of the nuclear installation.

The characteristics and distribution of population in the region surrounding the site need to be analysed to evaluate and periodically update the potential impact of radioactive releases on the public during operational states as well as in accidental conditions. This analysis requires data on current and projected population densities and distributions in the region, including both resident population and transient population (see [II–1]).

The demographic information is also used when considering the feasibility of implementing an effective emergency response action plan for the nuclear installation site. Section 9 of SSG-92 [II–1] provides recommendations related to considering the feasibility of an effective emergency plan and the associated database on population distribution. In particular, this involves identifying populations in the site vicinity of nuclear installation that may be difficult to evacuate or shelter (e.g. vulnerable or sensitive populations such as in schools, prisons, hospitals or populations located in restricted settlement areas), as well as the public activities that may contribute to the emergency plan (hospitals, police services, firefighting services, municipal services), for activities associated with transport (see Section II–1.9). In addition, the feasibility of implementing an effective emergency plan, considering demography, can serve as a site selection criterion. This can be used either for excluding certain sites or for ranking potential sites, as discussed in Section 2.3.

The demography in the area around the site may be one of the non-safety related criteria for site selection, aiming, for example, at moving the nuclear installation away from the public, from some human induced activities or from some categories of vulnerable population.

Data on the population distribution and density is also necessary to evaluate human induced activities that could potentially impact operational safety. This includes making predictions about population distribution and growth, which may influence safety considerations throughout the projected lifetime of the nuclear installation.

### II–1.9. Industrial, military and transportation activities

Industrial, military and transportation activities are potential sources of hazards for a nuclear installation. Requirement 24 of SSR-1 [II–11] states that "**The hazards associated with human induced events on the site or in the region shall be evaluated.**" Human induced events may be either initiated by sources external to the site, potentially located far from the site, or from within the site's boundary. These sources are not directly related to the operational activities of the nuclear installation but are accidental in nature (see [II–14]).

The first step in constructing a database involves identifying potential sources of such human induced events, such as nuclear installations, industries and military facilities. This is also the case for land, river, sea or air transportation activities and related infrastructure (roads, railroads, airports, harbours, airways) (see [II–14]). These sources can be categorized as either stationary sources, such as installations that process, handle or store hazardous materials (e.g. flammable, explosive, toxic, corrosive or radioactive material), or mobile sources, including means of transportation of hazardous materials or potential projectiles, such as aircraft crashes. This database is partly already established through the description of the site's geography (see Section II–1.7).

Paragraph 2.19 of SSG-79 [II–14] states:

> "a prognosis should be made for possible regional development over the anticipated lifetime of the nuclear installation, with account taken of the degree of administrative control that could realistically be exercised over activities in the region. In this respect, allowance should be made for the fact that technologies in the chemical and petrochemical industries, as well as traffic densities, may evolve rapidly."

Attention also needs to be given to future human related activities that are currently in the planning stage, including land with potential for commercial development.

On the basis of the established list of nearby sources that could pose a hazard, potential human induced events from such activities need to be identified, including: air pressure wave; explosion; projectile impact; fire and heat; smoke and dust; flooding or lack of water; toxic gases, aerosols or liquids; and ground shaking or instability. Each potential event and its consequences need to be analysed, taking into account relevant factors including its magnitude, probability of occurrence and proximity to the site, as far as these can be readily determined. Some of these effects are significantly more important to safety than others, as they could impact both the installation's facilities and items important for safety. This includes the availability of evacuation routes, hindrance to the implementation of emergency procedures, as well as affecting access to the ultimate heat sink and external power grid.

The collected information is first utilized in a two-step screening process to exclude sources that need not be considered further, based on the screening distance values or screening probability values. A more detailed evaluation needs to be made for each type of source or event that cannot be excluded during the screening process. Recommendations on the evaluation of the consequences of human induced events are provided in SSG-79 [II–14].

Adapting the nuclear installation's design or its location within the site area may reduce the impact of the listed events and their consequence, provided that the effect of another hazard is not increased. As an example, burying the installation may reduce the impact of an airplane crash, but the benefit with regards to flooding hazard needs to be evaluated. As for other hazards, if the evaluated radiological consequences of a potential event are excessive and cannot be reliably prevented, mitigated or controlled to reduce this impact, the site might be deemed unsuitable. The feasibility of implementing an emergency action plan due to potential human induced events can also be a site selection criterion (see Section 2.3) for exclusion or for ranking the potential candidate sites.

### II–1.10. Ambient radioactivity and human health

The last and one of the most crucial steps in verifying site suitability involves evaluating the nuclear installation's potential effects on people and the environment. This requires collecting a comprehensive range of data on the site's characteristics, as mentioned above. Such an evaluation requires knowing the initial state of the site in terms of radioactivity and chemical toxicity, whether of natural origin or due to the presence of other industrial or nuclear activities.

Therefore, information on the ambient radiological conditions within the site area and within the local area and regional study area is needed. This includes information on the existing conditions, including an inventory of sources, their activity levels and their origin (natural or anthropogenic) for all applicable environmental media, including air, water bodies, aquatic sediments, plant or animal tissue, and soils and food (see section 3 of Ref. [II–1]). Determining the site's radiological background generally includes measurements of radiological parameters such total alpha and beta activity and gamma dose rates.

The definition of the initial state also includes the current health profiles of the communities that might be impacted by the nuclear activity, covering both local and regional areas.

## II–2. TOPICS PRIMARILY RELATED TO ENVIRONMENTAL AND SOCIOECONOMIC ASPECTS

### II–2.1. Aquatic and terrestrial environments

The aquatic environment includes the aquatic and wetland species (flora, fauna) and their habitats within the site and local study area and regional study area. The terrestrial environment includes flora and fauna, their habitats, corridors of wildlife and the soil.

One major use of data on the aquatic and terrestrial environments is to evaluate non-radiological and radiological impacts of the nuclear activities on the fauna and flora themselves but also on people (see Section II–1.10) through the food chain. In addition, the aquatic and terrestrial environment may provide significant resources, not only for the biodiversity they sustain but also for activities such as recreation, tourism and fishing. Consequently, the environment quality impacts not only the ecosystem but also has socioeconomic implications, contributing to the overall quality of life of people.

An environmental characteristics baseline on aquatic and terrestrial flora, fauna and food chain needs to be established for all of the above mentioned issues related to the radioactive releases and the non-radiological environmental impacts on aquatic and terrestrial environment, together with aquatic and terrestrial habitat and their associated biota. Two primary environmental factors that may affect the suitability of a nuclear installation site are the impact on ecosystem and the impact on key species. Therefore, documenting the biota at the proposed site needs to include descriptions of the terrestrial flora, bird communities, mammals, amphibians and reptiles, fish (including aquatic finfish, shellfish and their prey) and aquatic plants. The evaluation of potential impact includes any risk of destruction or adverse modification to their habitats, with a special focus on endangered species [II–4, II–12]. The baseline concentrations of radionuclide and hazardous substances in both aquatic and terrestrial biota, recorded prior to the construction of nuclear installation, are necessary to establish the existing levels of contamination.

The baseline also includes local human uses (see also Section II–2.2) if the local flora or fauna is consumed through private leisure (fishing, hunting, picking) or commercial activities.

Such baseline data support the evaluation of the nuclear installation's effects on the aquatic and terrestrial environments, notably the effects of nuclear and hazardous substances in the liquid and atmospheric effluents, effects of blasting and excavating, effects of impingement and entrainment and of the thermal plume, if any [II–4].

The data on the distribution of the species, on their vulnerability, together with the preliminary results of the evaluations of the consequences, need to be detailed enough to adapt the nuclear installation's design, particularly its location in the site or its depth, to minimize the consequences to the environment.

This environmental baseline includes the description of migratory and protected species that hold cultural, commercial or ecological significance. Additionally, it needs to include details about the site's ecosystem, highlighting areas subject to various conservation and protection standards. This involves collecting all available information on the protected and conservation areas established by federal, provincial and municipal jurisdictions (wilderness areas, parks, sites of historical or ecological significance, nature reserves, federal migratory bird sanctuaries, wildlife management areas). The existence of such protected and conservation areas near the site should be considered in the site selection criteria for ranking or exclusion (see [II–10]).

The environment's effect on the nuclear installation safety through biological hazards need to be examined. For that, various kinds of data are necessary, depending both on the site characteristics and the facility design. Biological phenomena, such as those involving organisms that could affect cooling water systems, need particular attention. Additionally, biological hazards posed by rodents and birds could potentially damage or interfere with electrical systems and other critical components within the installation.

### II–2.2. Land uses

The information on land use is a major factor to determine site suitability from health, safety and environmental viewpoints. The information on baseline land use is essential for predicting how public utilization of land and water resources in the surrounding area may impact the safety of the nuclear installation. This is particularly important because anthropogenic activities can alter natural hydrological and geotechnical hazards (see [II–5]). Examples are provided in para. A.35 of SSG-35 [II–10], including "Land use activities such as those that influence water courses or the stability of slopes affecting the site, such as upstream dams, major users of river water abstraction and industries that could deposit large amounts of debris into a river upstream of the site."

Requirement 27 of SSR-1 [II–8] states that "**The uses of land and water shall be characterized in order to assess the potential effects of the nuclear installation on the region.**" To meet this requirement, specific data are needed on food and water use, such as: location of the consumers (local, regional, spread in the national territory, abroad), quantities consumed, food processing, as well as details on the natural food chain up to the consumed species. Data on land and water use may also be necessary for evaluating the potential negative socioeconomic impacts of the nuclear installation on surrounding communities and the region. This includes assessing how the construction and operation of the nuclear installation might affect the population's ability to continue utilizing these resources. Paragraph 3.28 of SSG-92 [II–1] provides an example of how current uses might be impacted by alterations in water temperature and discharge of radioactive material from the nuclear installation. Other potential uses such as historic, archaeological or cultural resources need to be considered [II–12]. Reference [II–4] indicates that baseline land use information is essential for determining the appropriate size of the proposed exclusion zone for the site.

Data are also needed on the location, nature and extent of water usage including drinking water, uses of water for recreational purposes, for irrigation and fishing, as well as on the land devoted to agricultural uses, farming, hunting, industrial, commercial, institutional and recreational purposes, and any natural resource extraction (see section 3 of Ref. [II–1] for more details). Additionally, expected changes in land and water use in the future must be considered in the assessment.

The distribution of the local population needs to be assessed to ensure that their ability to continue land use activities within the emergency planning zone does not impede implementation of the emergency response plan. This evaluation needs to be periodically reviewed throughout the installation's lifetime.

Depending on the national regulations, land use and land use planning may be included as site selection criteria (see Section 2.3). Section 3 of SSG-92 [II–1] provides recommendation on investigations necessary on land and water uses in the region surrounding the site.

### II–2.3. Other societal or economic aspects

Other societal information might be necessary during the site selection and characterization process. For example, Ref. [II–12] notably mentions aspects related to public acceptance, aesthetics and environmental justice considerations. Reference [II–15] provides guidance on public participation (with a special focus on Indigenous participation and involvement and assessment of the impact on their rights), as well as on health, social, cultural and economic effect (on living and working conditions).

It is necessary to identify and evaluate potential concerns of the communities surrounding the proposed site throughout the siting process, notably through dialogue with the public. Reference [II–12] states:

> "Concerns that could be a factor include public resistance to the use of nuclear power, fear that an environmental resource or culturally sensitive area would be unduly compromised, the potential disproportionate impact to minority and/or low income populations, or preference to economic growth opportunities based on occupations, such as tourism, and not those related to industry or power generation."

Potential environmental justice considerations have to be considered in case of proximity of identified low income and minority populations to the site [II–12]. Reference [II–16] recommends that the siting decisions reflect fair treatment and meaningful involvement of all people, regardless of race, ethnicity, culture, income or educational level to ensure equitable consideration and to minimize disproportionate impacts on low-income populations and minorities.

Reference [II–12] provides examples of aesthetic impacts during the construction phase (e.g. construction noise, odours, dust, at the plant and along transmission and access corridors) and aesthetic perception created by the plant layout during the operation phase (e.g. landscaping, architectural features).

Additional considerations should be given to land availability, cost, factors affecting site constructability (such as availability to deliver an adequate and skilled workforce or construction materials on the site, or the ability to bring electricity to the transmission grid) and, for brownfield sites, the need for decontamination and cleanup. Further details are provided in Ref. [II–12], particularly appendix C-4, where these aspects are discussed in depth.

### References to Annex II


[II–1] INTERNATIONAL ATOMIC ENERGY AGENCY, Investigation of Site Characteristics and Evaluation of Radiation Risks to the Public and the Environment in Site Evaluation for Nuclear Installations, SSG-92, revision to IAEA Safety Standards Series No. NS-G-3.2, IAEA, Vienna, (in preparation).

[II–2] INTERNATIONAL ATOMIC ENERGY AGENCY, Seismic Hazards in Site Evaluation for Nuclear Installations, IAEA Safety Standards Series No. SSG-9 (Rev. 1), IAEA, Vienna (2022).

[II–3] INTERNATIONAL ATOMIC ENERGY AGENCY, Geotechnical Aspects in Siting and Design of Nuclear Installations, SSG-93, revision to IAEA Safety Standards Series No. NS-G-3.6, IAEA, Vienna (in preparation).

[II–4] CANADIAN NUCLEAR SAFETY COMMISSION, Reactor Facilities - Site Evaluation and Site Preparation for New Reactor Facilities, REGDOC-1.1.1, CNSC, Ottawa, ON (2018).

[II–5] INTERNATIONAL ATOMIC ENERGY AGENCY, Meteorological and Hydrological Hazards in Site Evaluation for Nuclear Installations, IAEA Safety Standards Series No. SSG-18, IAEA, Vienna (2011).

[II–6] INTERNATIONAL ATOMIC ENERGY AGENCY, Managing Siting Activities for Nuclear Power Plants, IAEA Nuclear Energy Series No. NG-T-3.7 (Rev. 1), IAEA, Vienna (2020).

[II–7] INTERNATIONAL ATOMIC ENERGY AGENCY, Format and Contents of the Safety Analysis Report for Nuclear Power Plants, IAEA Safety Standards Series No. SSG-61, IAEA, Vienna (2021).

[II–8] INTERNATIONAL ATOMIC ENERGY AGENCY, Site Evaluation for Nuclear Installations, IAEA Safety Standards Series No. SSR-1, IAEA, Vienna (2019).

[II–9] INTERNATIONAL ATOMIC ENERGY AGENCY, Seismic Design for Nuclear Installations, IAEA Safety Standards Series No. SSG-67, IAEA, Vienna (2021).

[II–10] INTERNATIONAL ATOMIC ENERGY AGENCY, Site Survey and Site Selection for Nuclear Installations, IAEA Safety Standards Series No. SSG-35, IAEA, Vienna (2015).

[II–11] INTERNATIONAL ATOMIC ENERGY AGENCY, Volcanic Hazards in Site Evaluation for Nuclear Installations, IAEA Safety Standards Series No. SSG-21, IAEA, Vienna (2012).

[II–12] IDAHO NATIONAL LABORATORY, Site Suitability and Hazard Assessment Guide for Small Modular Reactors, Rep. INL/EXT-13-29749, INL, Idaho, IL (2013).

[II–13] EUROPEAN ATOMIC ENERGY COMMUNITY, FOOD AND AGRICULTURE ORGANIZATION OF THE UNITED NATIONS, INTERNATIONAL ATOMIC ENERGY AGENCY, INTERNATIONAL LABOUR ORGANIZATION, INTERNATIONAL MARITIME ORGANIZATION, OECD NUCLEAR ENERGY AGENCY, PAN AMERICAN HEALTH ORGANIZATION, UNITED NATIONS ENVIRONMENT PROGRAMME, WORLD HEALTH ORGANIZATION, Fundamental Safety Principles, IAEA Safety Standards Series No. SF-1, IAEA, Vienna (2006), https://doi.org/10.61092/iaea.hmxn-vw0a.

[II–14] INTERNATIONAL ATOMIC ENERGY AGENCY, Hazards Associated with Human Induced External Events in Site Evaluation for Nuclear Installations, IAEA Safety Standards Series No. SSG-79, IAEA, Vienna (2023).

[II–15] GOVERNMENT OF CANADA, Practitioner's Guide to Federal Impact Assessments (2020), https://www.canada.ca/en/impact-assessment-agency/services/policy-guidance/practitioners-guide-impact-assessment-act.html.

[II–16] NUCLEAR REGULATORY COMMISSION, General Site Suitability Criteria for Nuclear Power Stations, Regulatory Guide 4.7, NRC, Washington, DC (2014).

## Annex III

## TENNESSEE VALLEY AUTHORITY CLINCH RIVER NUCLEAR SITE: SITE SELECTION AND EVALUATION FOR SMALL MODULAR REACTORS

This annex describes the process that the Tennessee Valley Authority (TVA) used to identify proposed and alternative sites in its Clinch River Nuclear (CRN) Site Early Site Permit (ESP) application for two or more small modular reactors (SMRs). The site selection process consists of an initial identification of a Region of Interest (ROI) for the siting of a new nuclear power plant, selection and evaluation of suitable potential sites, and a subsequent comparative analysis of candidate sites to identify the proposed and alternative sites detailed in the CRN Site ESP application. The TVA site selection process was performed consistent with established industry guidance.

### III–1. INTRODUCTION

The TVA is a corporate agency of the United States of America that provides electricity for business customers and local power distributors serving nine million people in parts of seven states in the south-eastern United States of America. In accordance with the TVA Act of 1933, the agency is committed to maintaining a leadership role in innovation. In accordance with this mission, TVA works to develop innovative technologies that will meet future electricity demand [III–1].

This case study describes the process that TVA used to identify proposed and alternative sites in its CRN Site ESP application. The site selection process complied with United States Nuclear Regulatory Commission (NRC) regulations and guidance. This case study does not address ongoing or future work being performed or considered by TVA with respect to SMR technologies or the CRN Site.

Early in 2010, TVA determined focus areas for potential future technology innovations within the agency. Based on TVA's strategy and environmental policy to focus on cleaner energy, and the potential for SMR technology to make a powerful and beneficial impact in the Tennessee Valley, it was determined that the deployment of SMRs need to be among TVA's top choices for focused innovation efforts. As a result, TVA began evaluating the feasibility of deploying SMRs within its service area.

A key characteristic of SMRs that contributed to TVA's interest in deploying the technology was the substantial use of modular construction, which would enable major portions of the plant to be fabricated in controlled manufacturing environments and shipped to the plant location via rail, trucks and/or barges. In addition to the improved construction techniques, other advantages identified by TVA included the anticipated ability to support TVA's strategic vision of cleaner power generation, the ability to incrementally add generation when and where it is needed, and the anticipated ability to re-power retiring fossil-generation sites, thereby taking advantage of existing assets (e.g. environmental permits, transmission capabilities, water availability).

As TVA continued to evaluate SMRs, it became increasingly apparent that SMRs could also support TVA's desire for a more diversified generating fleet; satisfy an increasing need for continuous, non-carbon emitting, low-cost energy generation; provide for a next-generation nuclear technology that is more flexible and has increased safety margins; and could help TVA and the United States of America improve energy security.

In 2014, TVA decided to pursue an ESP application using a plant parameter envelope approach, a process authorized under 10 Code of Federal Regulations Part 52 [III–2]. The plant parameter envelope is built by creating a set of bounding parameters to capture the potential reactors to be considered, such as the maximum power output, heat rejected to the environment, land area disturbed. The plant parameter envelope approach allows an applicant to determine the suitability of the site for nuclear power plant construction and operation without committing to a specific technology. During the period the applicant and the regulatory body are evaluating the site, developers continue to make progress on maturing their new reactor designs, enabling the applicant to make a better future technology selection decision.

Regulations in the United States of America provide for a typical plume exposure pathway emergency planning zone (EPZ) size of about 10 miles (16 km) surrounding a commercial nuclear power plant. The enhanced safety inherent in SMR designs (e.g. passive safety designs, increased safety margin, reduced risk, smaller and slower fission product accident releases, and reduced potential for dose consequences to the populace in the vicinity of the plant) allows for the potential to site SMRs closer to population centres than traditional large light water reactors (LLWRs), while utilizing the same acceptance criteria found in NUREG-0396 [III–3] that current LLWRs use for their 10 mile EPZ. Applying this same dose-based, consequence-oriented approach results in scalable EPZs specific to a given design, which enables TVA to consider the establishment of smaller EPZs (potentially confined within site boundaries), which could result in increased siting options.

In its ESP application, TVA included a dose-based, consequence-oriented methodology to determine an appropriate EPZ size for SMRs. TVA also included two emergency plans and two sets of exemptions for NRC to evaluate during its review of the ESP application. The NRC approved the EPZ sizing methodology that TVA proposed for calculating potential dose and approved both emergency plans and sets of exemptions. This enables TVA to implement the EPZ sizing methodology and propose an appropriate EPZ size in a future licensing action based upon a specific reactor design. As part of this future licensing action, the NRC would evaluate TVA's implementation of the EPZ sizing criteria using the methodology approved during the ESP for calculating potential dose to the public from postulated events.

The NRC recognizes that existing emergency preparedness regulations are primarily focused on LLWRs and non-power reactors. To better address the possible future deployment of SMRs, the NRC is proceeding with the development of new emergency preparedness requirements that adopt a consequence oriented, risk informed, performance-based and technology-inclusive approach that could be implemented by SMR licensees and would continue to provide reasonable assurance that adequate protective measures can and will be implemented to protect the health and safety of the public [III–4]. This approach is like that proposed by TVA in its ESP application.

TVA, as authorized by the TVA Act [III–5], entered into an agreement with the US Department of Energy to collaborate, and explore various options under which TVA could perform siting, engineering and licensing for potential construction and operation of SMRs to demonstrate the capability of the technology. In part, TVA proposed SMR technologies to supply federal mission critical customers with reliable electricity from sources that are less vulnerable to supply disruption than typical commercial nuclear power generation facilities [III–1].

## III–2. SITE SELECTION AND THE EARLY SITE PERMIT APPLICATION

This case study describes the process TVA used to select candidate sites and subsequently identify proposed and alternative sites for more detailed analysis of site suitability to demonstrate the capability of the SMR technology.

TVA's site selection process began with the identification of an ROI. Within that ROI, TVA used screening criteria to establish candidate areas, potential sites and candidate sites, resulting in the selection of proposed and alternative sites. As a result of the site selection process, TVA ultimately selected the CRN site as the preferred location to potentially demonstrate the capability of the SMR technology [III–1]. The site selection process was conducted in accordance with the overall process outlined in Ref. [III–6].

At the time TVA was developing the ESP application, its business objectives included:

— Using SMR technology to enhance nuclear safety and security;
— Demonstrating the ability to deploy SMRs incrementally to efficiently meet demand;
— Using SMR technology to assist federal facilities with meeting carbon reduction objectives;
— Demonstrating that SMR technology is capable of supplying federal mission critical loads with reliable power from generation and transmission that is less vulnerable to supply disruption from intentional destructive acts and natural phenomenon than typical commercial nuclear power generation facilities and transmission systems [III–7].

In May 2016, TVA submitted an application for an ESP for SMR units at the CRN site. In accordance with NRC regulations [III–2], TVA's ESP application included an Environmental Report analysing the environmental effects of construction, operation and decommissioning of two or more SMRs. The Environmental Report included an overview of the site selection process and identified and evaluated a set of alternative sites. The purpose of the evaluation was to verify that a reasonable suite of candidate sites had been considered, and to demonstrate that there was no 'obviously superior' site for the potential construction and operation of the SMRs [III–1].

As part of its overall review of TVA's ESP application, the NRC also analysed the safety characteristics of the proposed site and applicable emergency planning information.

Selection of a proposed site was a necessary early step in the development of the ESP application. Because TVA had not selected a specific SMR technology prior to initiating development of the ESP application, a plant parameter envelope approach, which identified bounding parameters for a surrogate plant, was used during the site selection process to evaluate the suitability for construction and operation [III–1].

In response to TVA's ESP application, the NRC prepared an environmental impact statement (EIS) using information from the Environmental Report, the CRN Site Safety Analysis Report and supplemental information provided by TVA during the NRC's review. Development of the EIS by the NRC satisfied the US National Environmental Policy Act requirements for federal agencies. NRC regulations require that the EIS be prepared in conjunction with an application for an ESP and include an evaluation of alternatives to the proposed action, in accordance with 10 Code of Federal Regulations Part 51 [III–8], including consideration of alternative sites [III–9].

## III–3. SITE SELECTION PROCESS

TVA's business objectives served as the initial basis for the site selection process. TVA's site selection process was conducted in accordance with the overall process outlined in the Electric Power Research Institute (EPRI) Siting Guide [III–6] and consisted of 1) identifying a ROI for siting SMRs; 2) identifying suitable candidate areas (i.e. elimination of those areas deemed not suitable for siting a SMR) within the ROI; 3) canvassing candidate areas for suitable potential sites; 4) evaluating suitable potential sites; and 5) performing a comparative analysis of candidate sites and identifying proposed and alternative sites [III–7].

Figure III–1, TVA Site Selection Process, summarizes the process used by TVA. Each step in the process is summarized in the following paragraphs.

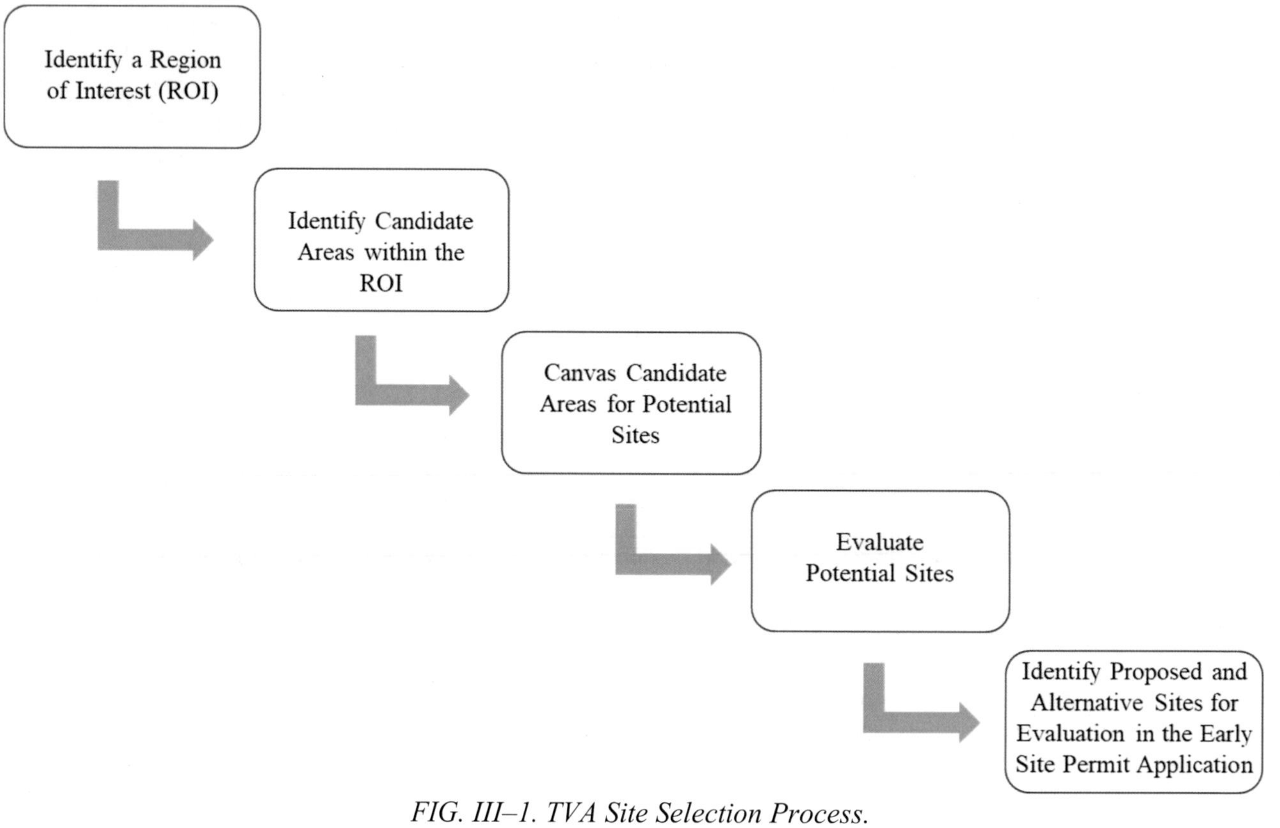


*FIG. III–1. TVA Site Selection Process.*

### III–3.1. Region of interest

A significant implication of the previously identified objectives is that TVA or its customer control the means to physically access the plant, including critical infrastructure to operate the plant [III–7]. As a result, the TVA Power Service Area was identified as the ROI for the SMR project. TVA's ROI is depicted in Fig. III–2.

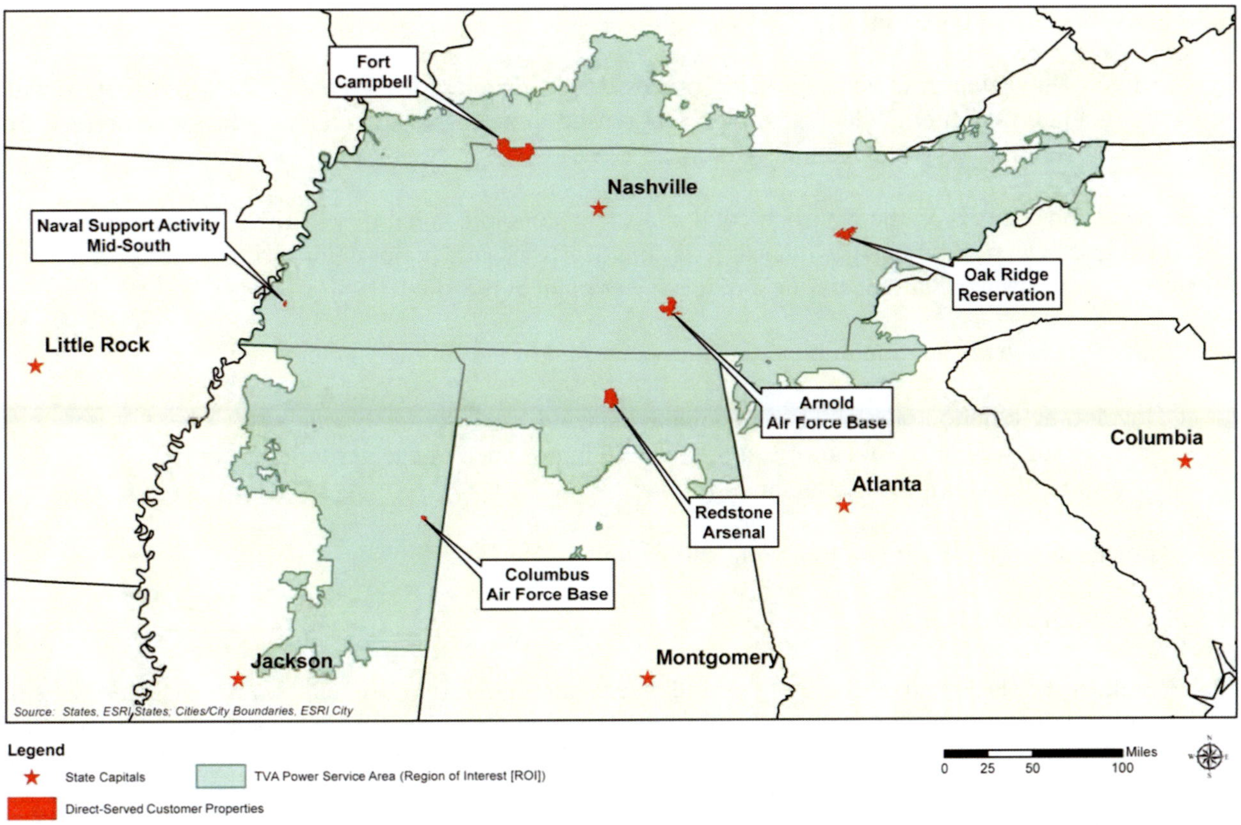


*FIG. III–2. TVA Region of Interest (reproduced from Ref. [III–1] with permission).*

### III–3.2. Identification of candidate areas

Following identification of the ROI for the SMR project, the ROI was screened to identify candidate areas. TVA identified candidate areas within the ROI by using geographical information systems (GIS) analyses of the entire ROI and applying exclusionary criteria, consistent with the exclusionary and avoidance criteria identified in the EPRI Siting Guide [III–6], to eliminate areas that were either considered to be not suitable — or significantly less suitable — than other potential areas for siting an SMR. The exclusionary criteria were based on the objectives of the project. The regional screening process included safety and security considerations, proximity to federal customers, distance to viable water sources, and other physical features that provided insights into site suitability within the ROI [III–7].

### III–3.3. Identification of potential sites

Investigations conducted as part of the site selection process were executed using existing reconnaissance level data available in the public domain or in TVA files. TVA consulted with its federal customers to identify potential sites within the previously identified candidate areas. TVA-owned properties adjacent to the candidate areas were also included for selection of potential sites. Potential sites were evaluated based on consistency with TVA objectives for the project, with consideration given to land use plans and other requirements associated with existing missions and activities.

Each candidate area was examined to identify locations that would provide sufficient land for the suitable arrangement of two or more SMRs, other necessary facilities and a reasonable site boundary. Principal considerations included land availability, proximity to a water source, presence of sensitive resources (e.g. wetlands), existing transmission lines, existing transportation infrastructure, obvious topographic concerns and flexibility to optimize site layout and design for environmental and cost mitigation purposes [III–9].

### III–3.4. Evaluation of potential sites

The objective of this step in the site selection process was to subject the potential sites to a more detailed evaluation of suitability criteria. General siting criteria used to evaluate the potential sites were derived from those presented in the EPRI Siting Guide [III–6].

TVA's evaluation focused on the environmental and socioeconomic suitability of the sites, health and safety issues considered when making major nuclear licensing decisions, and major engineering issues related to cost or difficulty of constructing and operating a new nuclear plant at the site [III–5].

The overall process for evaluation of potential sites consisted of the following elements:

- Developing criterion ratings for each potential site;
- Establishing weight factors to reflect the relative importance of each criterion;
- Determining composite site suitability ratings.

These elements are discussed in the following subsections.

#### *III–3.4.1. Criterion Ratings*

Each potential site was assigned a rating of 1–5 (1 = least suitable, 5 = most suitable) for each general siting criterion. Information sources for these evaluations included publicly available data, data available from TVA files and personnel, United States Geological Survey topographic maps, and satellite images. Reconnaissance level information was used in the evaluations.

#### *III–3.4.2. Weight Factors*

Weight factors reflecting the relative importance of these criteria were developed by a multi-disciplinary committee in the areas of nuclear power plant site suitability. This committee was composed of subject matter experts in geology/geotechnical/seismic, hydrology, ecology and wetlands, land use/zoning, transmission, socioeconomics, engineering, and public/institutional relations. The weight factors were derived using methodology consistent with the EPRI Siting Guide [III–6]. Weight factors were assigned a rating of 1–10 (1 = least important, 10 = most important).

#### *III–3.4.3. Composite Suitability Ratings*

Ratings reflecting the overall suitability of each site were developed by multiplying criterion ratings by the weight factors and summing over all criteria for each site.

The TVA evaluation of the remaining potential sites included field reconnaissance site visits to each site. TVA considered the impacts of offsite corridors for water supply pipelines, transmission lines, rail lines and primary roads. TVA used the total weighted scores for each site to determine the most suitable candidate sites [III–7].

## III–4. IDENTIFICATION OF PROPOSED AND ALTERNATIVE SITES

The evaluation of potential sites resulted in the selection of a smaller set of candidate sites (the CRN site and three alternative sites). The TVA Environmental Report included in the ESP application described the environmental aspects of the proposed site (i.e. the CRN site) in detail. In addition, the three alternative sites were examined in the ESP application relative to the following subject areas: land use, water resources, terrestrial and aquatic ecology, socioeconomics and environmental justice, cultural resources and historic properties, air quality, non-radiological health, radiological health and postulated accidents [III–9]. TVA's proposed and alternative sites are depicted in Figures III–3 and III–4.

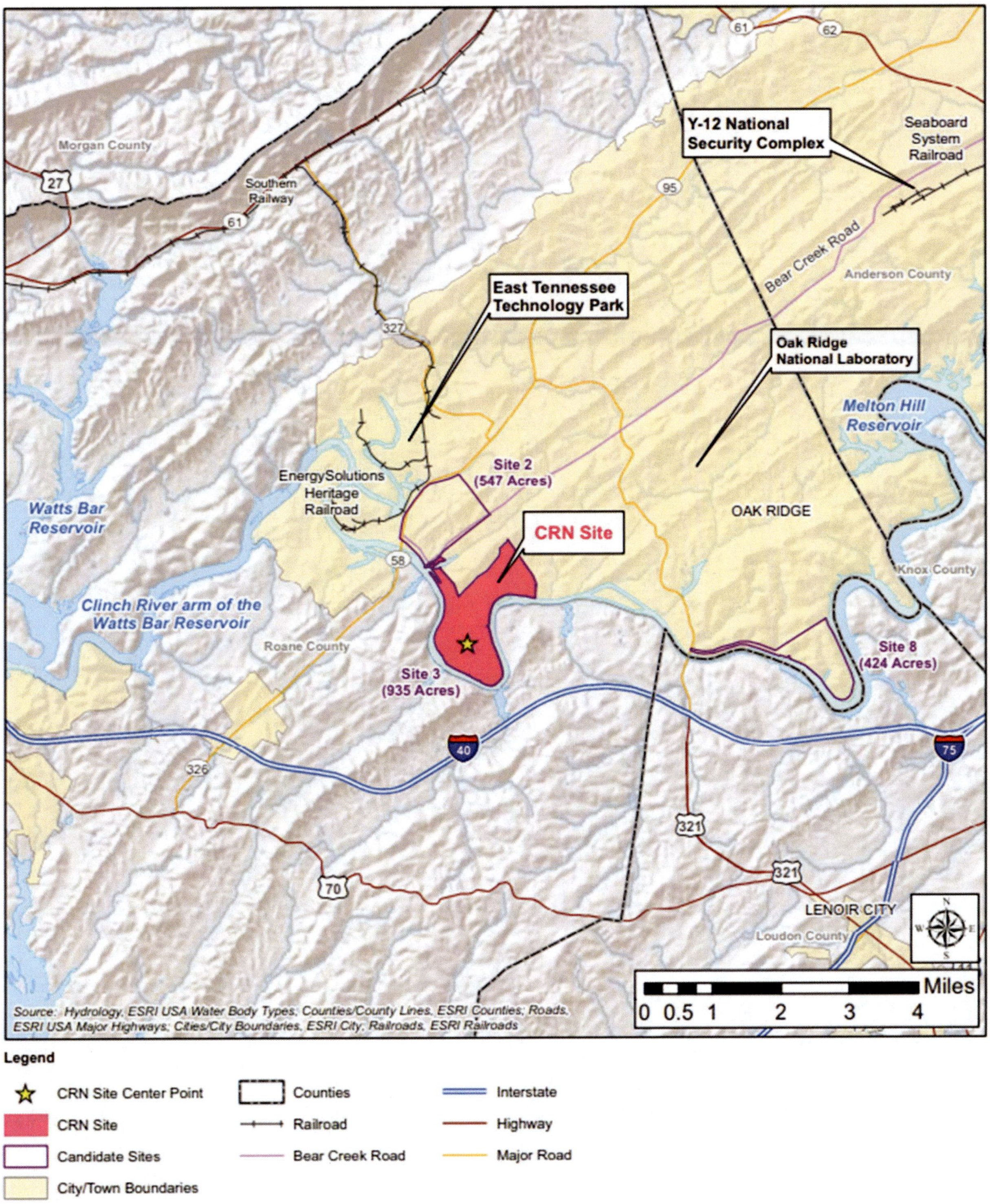


*FIG. III–3. Oak Ridge Reservation Area candidate sites (reproduced from Ref. [III–1] with permission).*

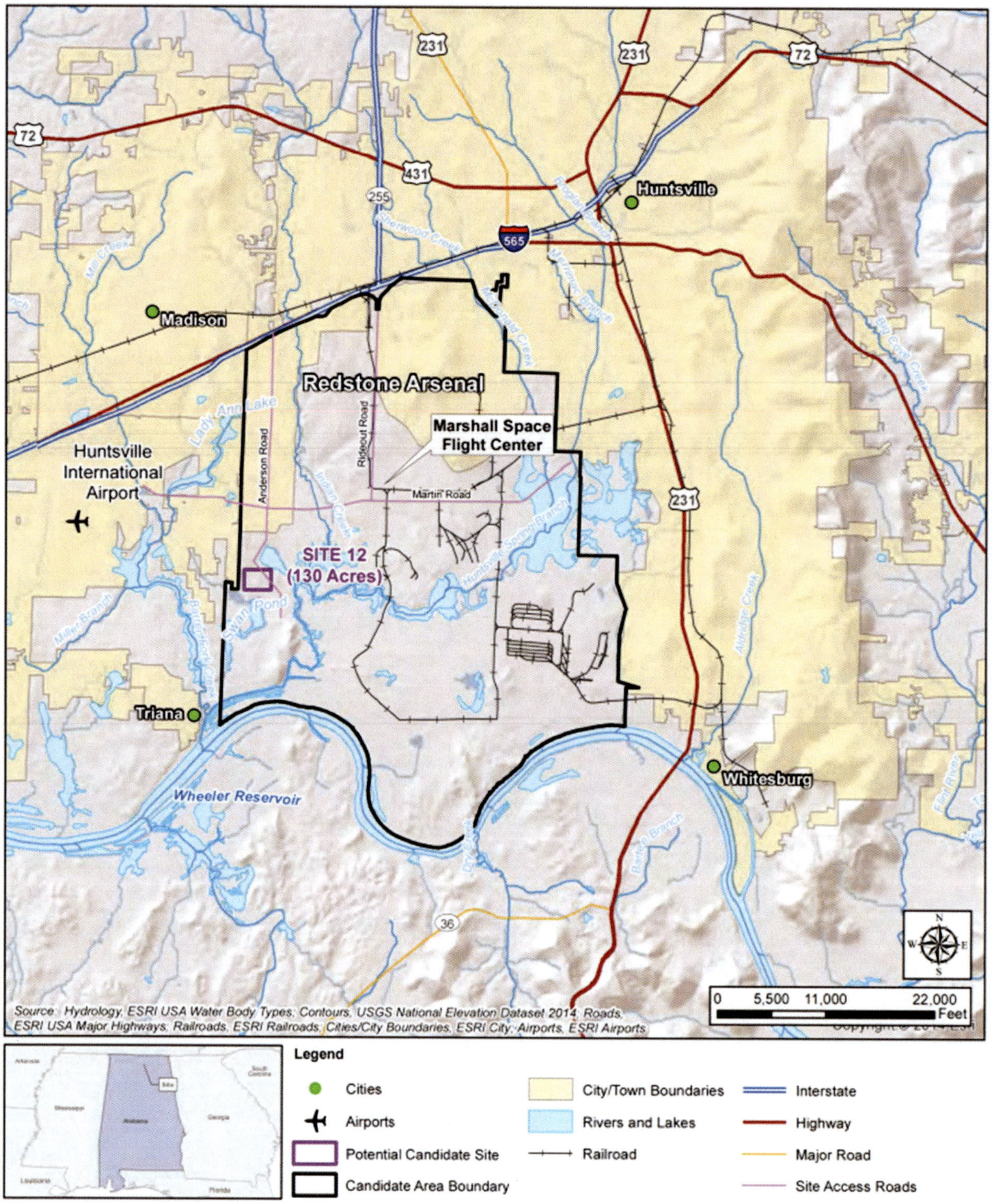


*FIG. III–4. Redstone Arsenal candidate site (reproduced from Ref. [III–1] with permission).*

## III–5. TVA EVALUATION OF THE PROPOSED AND ALTERNATIVE SITES

TVA considered both environmental criteria and technical criteria in its scoring of the sites. However, regulatory guidance associated with siting considers only environmental factors in the comparison of the sites to determine whether any is environmentally preferable [III–9].

Based on a detailed environmental review of the CRN site and the three alternative sites, TVA determined that none of the sites were environmentally preferable. Therefore, the CRN site was selected as the preferred site for the following business reasons:

- The CRN site is managed by TVA and is designated in TVA's land use plan for TVA project operations, which includes power production. There were no cost or schedule impacts associated with land transfer or land use re-designation;
- The CRN site was previously considered to be a suitable location for the Clinch River Breeder Reactor Project (CRBRP). As part of the CRBRP, TVA had collected historical environmental data and conducted site related research which reduced the environmental acceptance risk compared to the alternative sites;
- The CRN site is significantly larger than the alternative sites, allowing for greater flexibility for selecting a location within the site for the placement of SMRs and ancillary facilities that avoids or minimizes environmental impacts, particularly impacts to natural areas and terrestrial ecology;
- A large portion of the CRN site was previously cleared and the ground was prepared for CRBRP activities. In addition, some infrastructure installed for the CRBRP and still in place at the CRN site can be utilized for the SMR project. Although the alternative sites are part of federal installations and may contain limited infrastructure and structures, each of the three were greenfield sites which would require significant clearing. Because none of the alternative sites were determined to be environmentally preferable to the proposed CRN site, TVA concluded that none of the alternative sites were obviously superior to the CRN Site [III–1].

## III–6. REGULATORY BODY EVALUATION OF THE TVA SITE SELECTION PROCESS

The NRC's review of alternative sites consists of a two-part sequential test. The first part of the test determines whether any of the alternative sites are environmentally preferable to the proposed site. To determine if a site is environmentally preferable, the NRC staff considers whether the applicant has (1) reasonably identified candidate sites, (2) evaluated the likely environmental impacts of the proposed action at these sites and (3) used a logical means of comparing sites that led to selection of the proposed site. Based on its independent review, the NRC staff determines whether any of the alternative sites are environmentally preferable to the applicant's proposed site. If the NRC staff determines that one or more alternative sites are environmentally preferable, it then proceeds with the second part of the test.

The second part of the test determines if an environmentally preferable alternative site is not simply marginally better but obviously superior to the proposed site. The NRC staff examines whether (1) one or more important aspects, either singly or in combination, of an acceptable and available alternative site are obviously superior to the corresponding aspects of the applicant's proposed site and (2) the alternative site does not have offsetting deficiencies in other important areas. Included in this part of the test is the consideration of estimated costs (i.e. environmental, economic, and time to construct the proposed plant) at the proposed site and at the environmentally preferable site or sites. An NRC staff conclusion that an alternative site is obviously superior to the applicant's proposed site would normally lead to a recommendation that the application for the permit be denied [III–9].

Based on its independent review conducted in accordance with part one of the sequential test, the NRC staff determined that none of the alternative sites were environmentally preferable to the proposed CRN site. Therefore, NRC staff concluded that none of the alternative sites would be obviously superior to the CRN site.

If the NRC staff had determined that one or more alternative sites were environmentally preferable, it would then have proceeded with the second part of the sequential test, described above. This second part of the test would have determined if an environmentally preferable alternative site is not simply marginally better but obviously superior to the proposed site.

Although NRC Guidance [III–10] considers only environmental factors in the comparison of the sites to determine whether any is environmentally preferable, TVA considered both environmental criteria and technical criteria in its scoring of the sites. Therefore, the NRC review team examined the TVA scores for each of the candidate sites based only on environmental issues and concluded that although the proposed CRN site did not receive the highest environmental score (including both ecological and socioeconomic factors), the scores of all the sites in the area were similar. The NRC review team also examined the values of the weighting factors that were applied by TVA to each of the site characteristics and concluded that no single one of these weighting factors was by itself sufficient to significantly skew the total score obtained by TVA for any one site or to alter the ranking for the top two sites. The review team concluded that TVA's consideration of non-environmental factors did not inappropriately affect the outcome of the TVA analysis. The NRC review team concluded that TVA's site selection process was reasonable, made full use of the available candidate site data and presented the data in a manner that permitted valid comparisons between the candidate sites. Overall, the review team concluded that TVA used a logical approach that adequately satisfied applicable NRC guidance for the identification of sites that are among the best in the ROI [III–9].

## III–7. OPERATIONAL EXPERIENCE / LESSONS LEARNED

TVA identified a number of opportunities for improving the site selection and evaluation process, including:

- Identifying high risk technical issues early in the site selection process (this may require a higher level of expertise);
- Identifying the importance of utilizing subject matter experts with experience implementing the siting process (preferably with local and/or regional expertise);
- Ensuring the site selection rating system gives proper weighting to those aspects that affect future project cost compared to those that have safety or environmental considerations;
- Ensuring the site selection process aligns with regulatory requirements and guidance used in licensing applications, or developing sufficient justification if the process deviates from guidance;
- Conducting regular meetings and ESP engagement with regulatory agency staff facilitates early identification and mitigation of potential deviations from guidance during the site selection process;
- Working closely with reactor vendor(s) to develop appropriate plant parameters envelope values;
- Determining, and consistently providing, the appropriate level of detail for inclusion in the site selection report (i.e. detailed vs. reconnaissance level data);
- Recognizing that there may be security related obstacles in obtaining data on or near federal installations;
- Ensuring a logical/consistent approach to evaluating all sites, including recognizing the potential for, and actively avoiding, bias that may come with the existence of detailed data associated with sites that have been previously evaluated (versus reconnaissance level data).

## References to Annex III


[III–1] TENNESSEE VALLEY AUTHORITY, Clinch River Nuclear Site, Early Site Permit Application, Part 3, Environmental Report, (Rev. 2), TVA, Knoxville, TN (2019).

[III–2] NUCLEAR REGULATORY COMMISSION, Licenses, Certifications, and Approvals for Nuclear Power Plants, 10 CFR 52, US Govt Printing Office, Washington, DC (2025).

[III–3] NUCLEAR REGULATORY COMMISSION, Planning Basis for the Development of State and Local Government Radiological Emergency Response Plans in Support of Light Water Nuclear Power Plants, Rep. NUREG-0396, NRC, Washington, DC (1978).

[III–4] NUCLEAR REGULATORY COMMISSION, Final Rule: Emergency Preparedness for Small Modular Reactors and Other New Technologies (RIN 3150-AJ68; NRC-2015-0225), SECY-22-0001, NRC, Washington, DC (2022).

[III–5] U. S. CODE, Title 16 Chapter 12A, Tennessee Valley Authority Act of 1933, Washington, DC (1933).

[III–6] ELECTRIC POWER RESEARCH INSTITUTE, Advanced Nuclear Technology: Site Selection and Evaluation Criteria for New Nuclear Power Generation Facilities, Rep. 3002005435, Palo Alto, CA. (2015).

[III–7] TENNESSEE VALLEY AUTHORITY, Tennessee Valley Authority Site Selection Report, TVA, Knoxville, TN (2016).

[III–8] NUCLEAR REGULATORY COMMISSION, Environmental Protection Regulations for Domestic Licensing and Related Regulatory Functions, 10 CFR 51, US Govt Printing Office, Washington, DC (2025).

[III–9] NUCLEAR REGULATORY COMMISSION, Environmental Impact Statement for an Early Site Permit (ESP) at the Clinch River Nuclear Site, Rep. NUREG-2226, NRC, Washington, DC (2019).

[III–10] NUCLEAR REGULATORY COMMISSION, Processing Applications for Early Site Permits, Review Standard RS-002, NRC, Washington, DC (2004).

## Annex IV

# EXAMPLES OF A GRADED APPROACH TO SITING FOR A RANGE OF NUCLEAR INSTALLATIONS

This annex describes the application of the graded approach to siting for a range of nuclear reactor facilities in Canada. In particular, it focuses on the scope and effort needed for reviewing and assessing site evaluation and site preparation activities for a variety of reactor technologies. This annex is based on information in annex III–2 of Ref. [IV–1].

## IV–1. GENERAL CONSIDERATIONS REGARDING A GRADED APPROACH

The IAEA Safety Glossary [IV–2] defines a graded approach as follows:

> "For a system of control, such as a regulatory system or a safety system, a process or method in which the stringency of the control measures and conditions to be applied is commensurate, to the extent practicable, with the likelihood and possible consequences of, and the level of risk associated with, a loss of control."

Reference [IV–1] on the application of a graded approach to the regulation of nuclear installations provides further elaboration on the graded approach, and section 3.1 states:

> "An example of a graded approach in general would be a structured method by means of which the stringency of application of requirements is varied in accordance with the circumstances, and the regulatory systems and the management systems used.
>
> "For example, a graded approach is a method in which:
>
> (1) The significance and complexity of a product or service are determined;
> (2) The potential impacts of the product or service on health, safety, security, the environment, and the achieving of quality and the organization's objectives are determined;
> (3) The consequences if a product fails or if a service is carried out incorrectly are taken into account.
>
> "The use of a graded approach is intended to ensure that the necessary levels of analysis, documentation and actions are commensurate with, for example, the magnitudes of any radiological hazards and non-radiological hazards, the nature and the particular characteristics of a nuclear installation, and the stage in the lifetime of the nuclear installation."

## IV–2. BACKGROUND ON SITING, ENVIRONMENTAL RISK ASSESSMENT AND SITING ACTIVITIES

Siting is the process to select an appropriate location for a nuclear installation. The siting process typically consists of two phases including site survey and selection. This is followed by site assessment, which concludes with a decision of whether a site is appropriate for the construction and operation of the nuclear installation, including derivation of the site related design basis.

An environmental risk assessment (ERA) is an important aspect of the siting process, in particular supporting site selection and assessment. An ERA identifies and characterizes the risks associated with contaminants, including nuclear or hazardous substances and other stressors within the environment.

Topics addressed in an ERA include atmospheric, surface, aquatic, geological, hydrological and terrestrial environment, background radioactivity and human health and non-human biota [IV–1].

As stated in Ref. [IV–1]

"in Canada, site preparation activities may include, but are not limited to:

— Construction of site access control measures;
— Clearing and grubbing of vegetation;
— Excavation and grading of the site to a finished elevation;
— Installation of services and utilities (domestic water, fire water, sewage, electrical, communications, natural gas) to service the future nuclear facility;
— Construction of administrative and support buildings inside the future protected area;
— Construction of environmental monitoring and mitigation systems; and
— Construction of flood protection and erosion control measures."

IV–3. CONSIDERATIONS

The following information is considered in application of the graded approach to siting:

— Descriptions of the reactor (e.g. power level, complexity, novelty);
— Environmental risk assessment criteria;
— Size of study area: the area potentially impacted by a postulated release to the environment during an accident;
— Site preparation activities and applicable criteria.

The technologies considered are a nuclear battery of about 600 kW, a high temperature gas reactor (HTGR) of about 10 MW and a traditional nuclear power plant (NPP) of about 600–1000 MWe.

Reference [IV–1] provides more detailed information on the technologies and criteria for the environmental risk assessment and site preparation activities. Typically, there are three study areas used in environmental risk assessments: site, local and regional, as described below (see Figure IV-1).

**Site study area:** The site study area is the project's footprint. In other words, it is where project activities would be undertaken, and it includes the project's proposed facilities, buildings and infrastructure.

**Local study area:** The local study area is defined as the area that exists outside the site study area boundary, where measurable changes to the environment may be anticipated as a result of the proposed activities at any phase of the project, either through normal activities or from possible accidents or malfunctions. The boundaries might change if appropriate following an assessment of the spatial extent of potential effects. The geographical boundary depends on the factor being considered (e.g. a local study area defined for the aquatic environment will differ from that defined for the atmospheric environment).

**Regional study area:** The regional study area is defined as the area within which the potential effects of this project may interact with the effects of other projects, resulting in potentially cumulative effects. The geographical boundary for the regional study areas is also specific to the factor being considered.

Considerations in determining a regional study area are provided in Section IV–4.4.

**Study Areas**

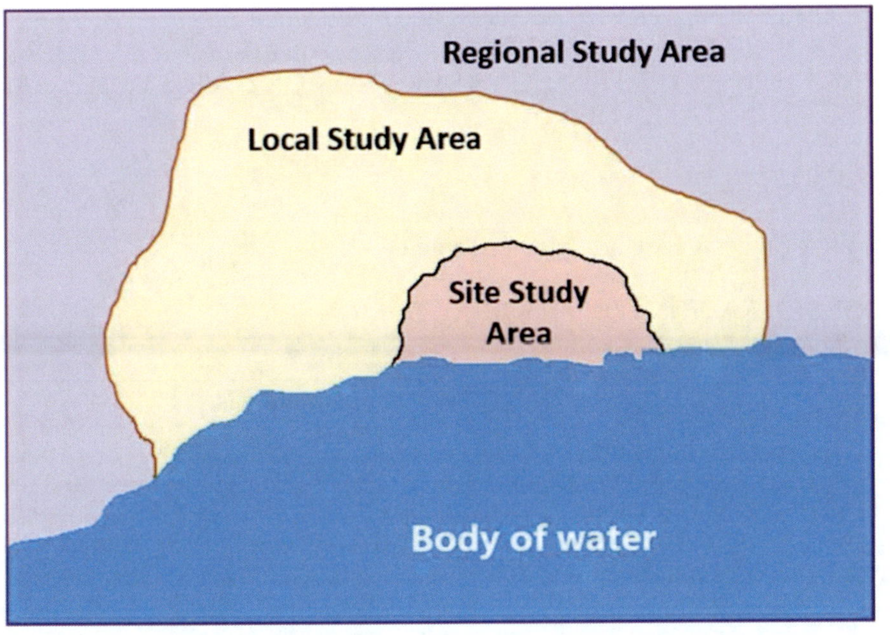


*FIG. IV–1. Types of study area (reproduced from Ref. [IV–1]).*

Considerations for the study areas, along with the focus and effort for the review and assessment of the three technologies taken from Ref. [IV–1] are presented in Table IV–1.

TABLE IV–1. SIZE OF STUDY AREAS FOR REACTOR TECHNOLOGIES: INFORMING ENVIRONMENTAL RISK ASSESSMENTS AND SITE EVALUATION
*(reproduced from Ref. [IV–1])*

| Factors | Nuclear battery (~ 2 MWth) | High temperature gas reactor (~ 10 MWth) | Large water cooled NPP (> 1000 MWth) |
|---|---|---|---|
| • Proposed footprint: layout of structures in the final layout state<br>• Excavation depth<br>• Projected release to the environment under postulated accident conditions | • Nuclear battery is expected to have a very small footprint<br>• Near-surface location<br>• Negligible release during accidents<br>• Study areas are small<br>• Limited amount of data to support site characterization would be needed<br>• Accident and malfunction analysis is straightforward due to simple design | • Small HTGR is expected to have a moderate size footprint<br>• May be located deep to address security issues by design<br>• Small release during accidents<br>• Study areas are relatively small<br>• Limited amount of data to support site characterization would be needed<br>• Accident and malfunction analysis is moderate due to a design that is of moderate complexity and has few process and safety systems | • NPP will have a large footprint<br>• Deep excavation<br>• Larger release during accidents<br>• Study areas are large<br>• Amount of data to support site characterization is extensive<br>• Accident and malfunction analysis is extensive due to a complex design and has several process and safety systems |

The regulatory focus and effort in site evaluation and the associated ERA are generally proportional to the size of the regional study area, with the primary consideration being the extent of the area (e.g. data to be collected) impacted by normal operation, as well as by accidents and malfunctions.

## IV–4. SITE PREPARATION ACTIVITIES: CONSIDERATION OF TOPICS IN THE SAFETY CASE FOR A NUCLEAR INSTALLATION

### IV–4.1. Physical design

Considerations for physical design are provided in Table IV–2.

TABLE IV–2. CONSIDERATIONS ON PHYSICAL DESIGN
*(reproduced from Ref. [IV–1])*

| Factors | Nuclear battery (~ 2 MWth) | High temperature gas reactor (~ 10 MWth) | Large water cooled NPP (> 1000 MWth) |
|---|---|---|---|
| • Design of any permanent feature that may support facility safety<br>• As pertaining to site preparation: information on the design measures such as flood protection and erosion control | • Nuclear battery is expected to have a very small footprint<br>• Review of design is simple due to very few, if any permanent features | • Small HTGR is expected to have a moderate size footprint, but may be located deep to address security issues by design<br>• Exclusion zone and emergency planning zones are likely to be relatively small<br>• Cooling needs are to be determined, but it is likely air-cooled<br>• Some earthworks may be necessary | • NPP will have a large footprint, with deep excavation<br>• Exclusion zone and emergency planning zones are likely to be relatively large (similar to that for the Darlington New Nuclear Project)<br>• Likely that access to two ultimate heat sinks is needed (a Fukushima-related requirement)<br>• Earthworks will be extensive |

### IV–4.2. Safety analysis

Considerations for safety analysis are provided in Table IV–3.

TABLE IV–3. CONSIDERATIONS ON SAFETY ANALYSIS
*(reproduced from Ref. [IV–1])*

| Factors | Nuclear battery (~ 2 MWth) | High temperature gas reactor (~ 10 MWth) | Large water cooled NPP (> 1000 MWth) |
|---|---|---|---|
| • Hazard Analysis<br>• Supports establishment of emergency planning zones & bounding envelope parameters | • Small number of postulated initiated events<br>• Deterministic safety analysis dominated by external events<br>• Probabilistic safety analysis complexity likely low but influenced by passive and inherent behaviours and simple design<br>• Severe accident analysis simpler<br>• Consequences very limited in area | • Small number of postulated initiated events<br>• Deterministic safety analysis dominated by external events<br>• Probabilistic safety analysis complexity likely low but influenced by passive and inherent behaviours as well as design complexity<br>• Severe accident analysis simpler<br>• Consequences very limited in area | • Significant complexity of postulated initiated events due to complex design<br>• Deterministic safety analysis needs to consider both external events and complex internal events<br>• Severe accident analysis is complex<br>• Consequences could be more wide ranging in area. |

### IV–4.3. Operating performance

Considerations for operating performance are provided in Table IV–4.

TABLE IV–4. CONSIDERATIONS ON OPERATING PERFORMANCE
*(reproduced from Ref. [IV–1])*

| Factors | Nuclear battery (~ 2 MWth) | High temperature gas reactor (~ 10 MWth) | Large water cooled NPP (> 1000 MWth) |
|---|---|---|---|
| • Operating performance includes an overall review of the conduct of the licenced activities and the activities that enable effective performance<br>• The area that site preparation activities are conducted on and extent of excavation are technology-specific<br>• Site preparation activities involve developing the foundation for the facility – it has to be done right | • Nuclear battery has a small footprint, so the cleared area is small and the excavation is not deep<br>• Site preparation activities are minimal and likely deferred to construction | • HTGR has a medium-sized footprint, so the cleared area is not extensive, but excavation may be deep (to address security-by-design)<br>• Site preparation activities are moderate – mainly due to excavation | • NPP has a large footprint, so the cleared area is large. Excavation is deep due to the size of the facility<br>• Site preparation activities are moderate |

The information in Tables IV–1 to IV–4 indicates that the following factors influence the focus and effort in the areas of physical design, safety analysis, and performing site preparation activities:

— Plant footprint and depth of excavation;
— Cooling technology;
— Complexity, novelty, potential harm and uncertainty in the reactor facility design.

### IV–4.4. Considerations for determining a regional study area

#### *IV–4.4.1. Physical design of reactor facility*

The Canadian Nuclear Safety Commission provides guides and regulatory documents that outline the information needed to support licence applications. These documents specify the level of detail required at each licensing phase and set forth requirements and guidance that influence the information an applicant must submit, particularly concerning the establishment of a regional study area. This area includes the regions potentially impacted by accidents, malfunctions and normal operation, ensuring that all possible environmental and safety impacts are thoroughly evaluated and documented.

Reactor vendors need to consider a broad range of applications, environments and regulatory requirements across all States of commercial interest when developing their designs. The process of establishing the regional study area begins with the nuclear installation design and the supporting data that justify safety-related claims. Vendors and applicants should ensure that their designs are robust enough to meet established safety objectives and protective action limits. This approach ensures that the designs are adaptable to various regulatory landscapes and capable of addressing safety considerations specific to each potential operating environment.

#### *IV–4.4.2. Postulated initiating events*

Postulated initiating events are hypothetical scenarios that could potentially lead to adverse effects on a nuclear installation, and they play a crucial role in the safety analysis. These events include internal occurrences — such as the failure of an installed component within the nuclear installation or an electrical fire — as well as external events, such as a significant earthquake, flooding or other natural hazards. By identifying and analysing these potential events, safety measures and mitigation strategies can be developed to ensure the continued safe operation of the nuclear installation under various adverse conditions.

#### *IV–4.4.3. Limiting credible accident and criteria for identification of the planning accident*

On the basis of the safety analysis, a list of limiting credible accidents is identified, focussing on scenarios that could significantly impact the nuclear installation. These scenarios include multiple and simultaneous severe external events that could potentially exceed the design basis, as outlined in Ref. [IV–3]. Applicants or licensees are expected to provide detailed information on these credible limiting accidents, including the associated source terms.

#### *IV–4.4.4. Source term and releases*

The source term is the list of radionuclides that could be released to the environment during a set of postulated accidents, taking into account the functioning of safety systems. The source term involves the duration of the release and parameters including the height at which a release may occur.

#### *IV–4.4.5. Meteorology for atmospheric dispersion and deposition models*

Site specific data related to meteorological characterization and modelling are utilized to assess the dispersion and deposition of isotopes released into the environment.

*IV–4.4.6. Dose assessment and distribution in consideration of pre-established dose criteria*

After developing dispersion and deposition models, exposure pathways are identified and doses are calculated. To establish the distances for which protective actions such as sheltering and evacuation may be necessary, the results are evaluated against dose criteria in order to plan the emergency response.

*IV–4.4.7. Regional study area determination process map*

Figure IV–2 shows the overall process used to determine the regional study area. The following external factors, and the distances for which protective actions would be needed, are considered when determining regional study areas:

- Scale and spatial extent of potential effects and impacts of the project;
- The physical location of potential receptors, including, where applicable, the movement patterns of potential receptors;
- Relationships between wildlife and vegetation;
- Community knowledge and indigenous knowledge;
- Current or traditional land and resource use by indigenous communities;
- Rights of indigenous peoples, including treaty lands, traditional territories and areas or sites used for cultural and spiritual practices;
- Physical, technical, ecological, social, health, economic and cultural considerations;
- Size, nature, location and known effects of past, present and foreseeable projects and activities, particularly for the regional study areas.

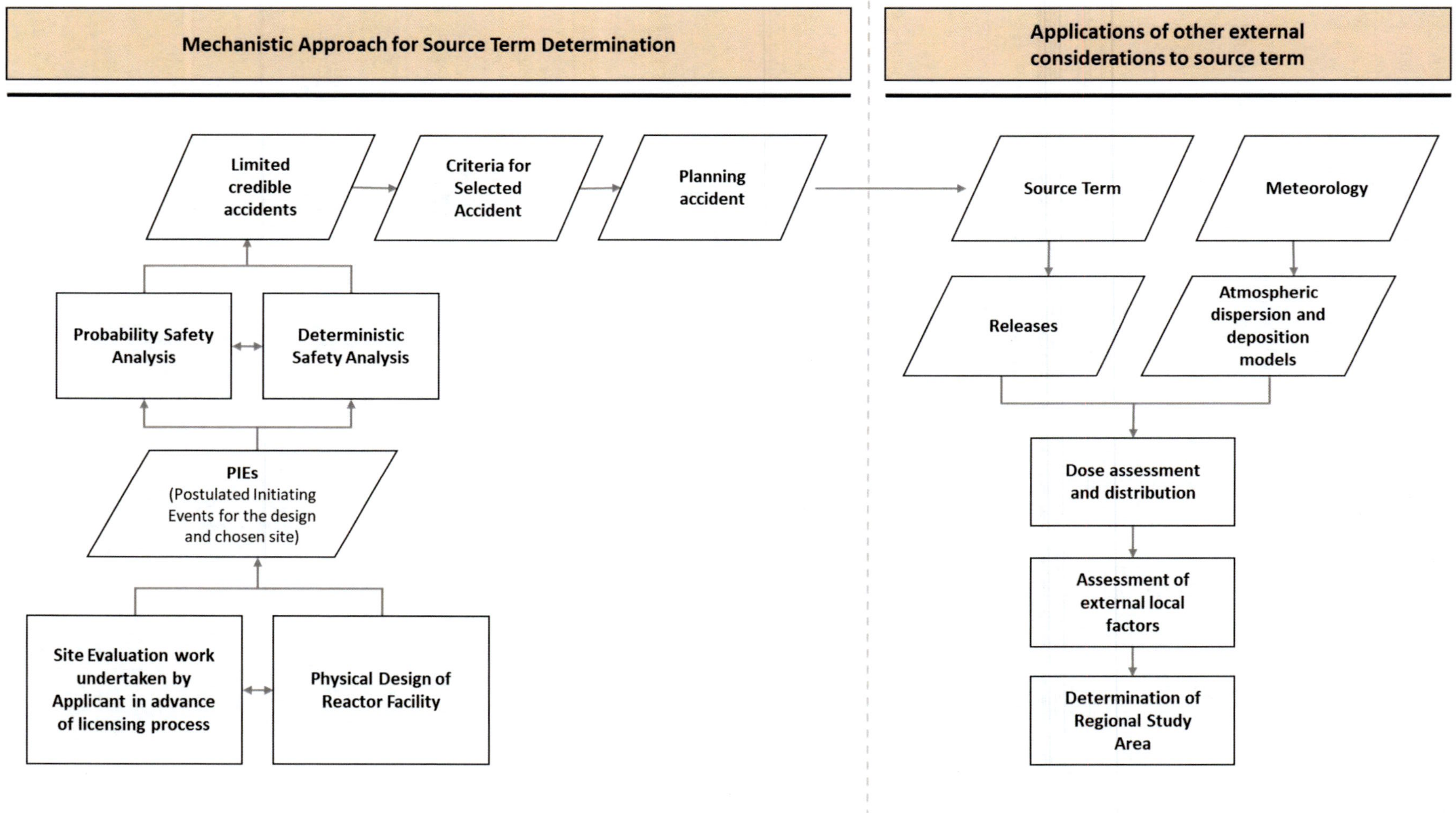


*FIG. IV–2. Overview of Canadian process for determining the regional study area.*

## References to Annex IV

[IV–1] INTERNATIONAL ATOMIC ENERGY AGENCY, Application of a Graded Approach in Regulating Nuclear Installations, IAEA-TECDOC-1980, IAEA, Vienna (2021).

[IV–2] INTERNATIONAL ATOMIC ENERGY AGENCY, Nuclear Safety and Security Glossary, Terminology Used in Nuclear Safety, Nuclear Security, Radiation Protection and Emergency Preparedness and Response, 2022 (Interim) Edition, IAEA, Vienna (2022).

[IV–3] CANADIAN NUCLEAR SAFETY COMMISSION, Nuclear Emergency Preparedness and Response, REGDOC-2.10.1, Version 2, CNSC, Ottawa, ON (2016).

## Annex V

## HIGH TEMPERATURE ENGINEERING TEST REACTOR: SEISMIC HAZARD ANALYSIS EXAMPLE FROM JAPAN

### V–1. INTRODUCTION

The Oarai Research and Development Institute of the Japan Atomic Energy Agency (JAEA) established the High Temperature Engineering Test Reactor (HTTR), which is first high temperature gas-cooled reactor (HTGR) in Japan [V–1]. The HTTR has superior safety features by using coated fuel particles, a graphite moderator and helium gas coolant. With the potential of supplying high temperature heat above 900°C, HTGR can be used not only for generation of power but also for process heat in several industrial fields. The long term high temperature operation (950°C/50days operation) was conducted by JAEA for demonstrating high temperature heat supply capability. In order to demonstrate the inherent safety features of HTGR, a loss of forced cooling test (at 30% power) was conducted in 2010. The severe accident was simulated by conducting a loss of forced cooling test during which the flow of reactor coolant was reduced to zero and the reactor scram was blocked. According to the result of the test, the reactor was able to be shut down and maintained in a stable condition without any management of operation. Through the process of designing, constructing and operating the HTTR, the JAEA successfully managed to accumulate useful data for developing the future commercial HTGR.

A large area of eastern Japan was affected by the 9.0 magnitude Great Pacific Coast Tohoku Earthquake that struck the Pacific Ocean off the coast of Sanriku on March 11, 2011. The epicentre of the great earthquake was located about 300 km away from the HTTR site at Oarai-machi, Ibaraki-prefecture. Due to maintenance of instruments and equipment and periodic inspection, the HTTR was shut down when the great earthquake occurred.

To monitor the earthquake response at the HTTR site, a system for seismic observation was installed [V–2]. Seismometers were installed at various depths around the foundation of the reactor building: to the north at depths of 250 m and 30 m; to the south at depths of 94 m, 30 m and 1 m; and to the west at depths of 174 m, 95 m, 32 m and 1 m. In addition, seismometers were installed inside the reactor building at several levels, including B3F, B1F, 1F and 2F, as well as embedded in the inner concrete at B3F and B1F. During the great earthquake, the maximum seismic acceleration recorded at the HTTR surpassed the design basis earthquake ground motion, which was established according to the regulatory guide issued in Japan in 1981. As a result, a visual inspection of the HTTR facility was conducted, accompanied by a seismic analysis using the recorded seismic acceleration to assess the integrity of the HTTR [V–3, V–4].

After the great earthquake, the nuclear regulatory authority in Japan published a seismic regulatory guide in 2013 [V–5]. Based on this guide, JAEA conducted seismic response analysis to confirm the integrity of structures, systems and components (SSCs) for the HTTR after the definition of design basis earthquake conditions. The same assessment methodology used for large-sized nuclear power plants was applied to the HTTR according to the request by the nuclear regulator in Japan.

### V–2. DESIGN BASIS EARTHQUAKE CONDITION

The first requirement of the regulatory guide is no on-site capable faults [V–6]. The capable faults are defined considering potential activities since the late Pleistocene (approx. 120 000–130 000 years ago). In case of no faults in the late Pleistocene, this guide additionally requires further geological investigation of potential capable faults for the middle Pleistocene (approx. 130 000–400 000 years ago). The second requirement is more precise methods to define design basis earthquake by survey of a three dimensional geological structure of the site assuming amplification of the ground motion due to the geological structure.

The HTTR site has no capable faults. Based on the regulatory requirements, JAEA has re-evaluated the design basis earthquake considering three types of earthquakes: inland earthquake, tectonic plate interface earthquake and ocean plate earthquake (slab earthquake) [V–7]. The regulatory guide requires an additional evaluation without specifying the epicentre, which is commonly applicable to the whole area in Japan. Figure V–1 shows evaluated results for the design basis earthquake (Ss). The evaluation without specifying the epicentre is enveloped by Ss-D. Using these spectra, acceleration waves can be obtained (see Fig. V–2). The value shown in each figure represents the peak ground acceleration. The maximum peak ground acceleration for the design basis earthquake is 973 cm/s$^2$ (in red in Fig. V–2), which was evaluated as Ss-1: earthquake at F3-F4 faults located east from the HTTR site. This is more than twice as large as the design basis earthquake defined previously (350 cm/s$^2$).

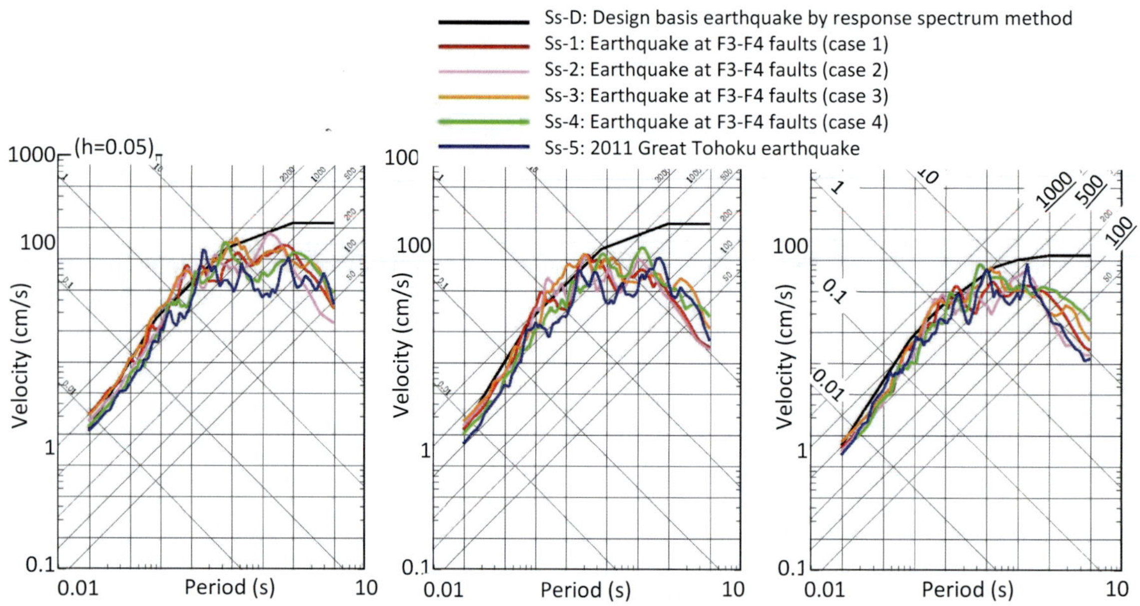


(a) NS horizontal (b) EW horizontal (c) vertical

*FIG. V–1. Spectra of design basis earthquake (reproduced from Ref. [V–7] with permission).*

| Design-basis earthquake | | Value: peak ground acceleration (cm/s$^2$) NS horizontal | EW horizontal | UD vertical |
|---|---|---|---|---|
| Ss-D | Design-basis earthquake by response spectrum method | 700 | | 500 |
| Ss-1 | Earthquake at F3-F4 faults (case 1) | 973 | 711 | 474 |
| Ss-2 | Earthquake at F3-F4 faults (case 2) | 835 | 761 | 436 |
| Ss-3 | Earthquake at F3-F4 faults (case 3) | 948 | 850 | 543 |
| Ss-4 | Earthquake at F3-F4 faults (case 4) | 740 | 630 | 405 |
| Ss-5 | 2011 Great Tohoku earthquake | 670 | 513 | 402 |

*FIG. V–2. Acceleration wave of design basis earthquake (reproduced from Ref. [V–7] with permission, vertical axis: acceleration (cm/s2), horizontal axis: time (s)).*

## V–3. SEISMIC RESPONSE ANALYSIS

Seismic response analysis for SSCs was performed according to the regulatory requirements. The typical results are shown in Table V–1. The evaluated results for all SSCs were less than the criteria, so that the integrity of SSCs in HTTR was confirmed under the newly evaluated design basis earthquake (see No. 1 in Table V–1).

The regulatory guide requires that the influence of the failure of low-class seismic components on high class components (called Class S in Japan) is examined. To confirm the influence, JAEA carried out the seismic response analysis for the low-class seismic components using the design basis earthquake. The evaluated results were less than the criteria, so that no influence on the Class S components was confirmed (see No. 2 in Table V–1).

The integrity of active components also needs to be assured. Therefore, JAEA evaluated the active function for isolation valves as a representative high class seismic component according to the regulatory requirements. As a result, the response acceleration generated by the design basis earthquake is less than the acceleration criterion under which the function was confirmed, so that the function of the active components was confirmed (see No. 3 in Table V–1).

TABLE V–1. TYPICAL RESULTS OF SEISMIC RESPONSE ANALYSIS
*(adapted from Ref. [V–7])*

| No. | Components | Evaluated stress/ acceleration | Criteria | Assessment |
|---|---|---|---|---|
| 1 | Reactor pressure vessel | 111 (MPa) | 350 (MPa) | OK |
| 2 | Overhead crane | 244 (MPa) | 279 (MPa) | OK |
| 3 | Isolation valve | 500 (cm/s$^2$) | 5880 (cm/s$^2$) | OK |

## V–4. CONCLUSION

The structural integrity of SSCs in the HTTR has been confirmed through the seismic response analysis after the recently defined design basis earthquake. Only the seismic assessment is described here because the most important external hazard in Japan is earthquakes. More recently, the regulatory requirements have been revised to include other external hazards, such as tsunami, tornado and volcanic eruption. Design basis external hazard conditions were defined against these external hazards. Under these conditions, JAEA has confirmed the integrity of SSCs and developed safety measures. These analyses could contribute to one of the case studies for siting evaluation.

**References to Annex V**


[V–1] INTERNATIONAL ATOMIC ENERGY AGENCY, Advances in Small Modular Reactor Technology Developments, 2020 Edition, A Supplement to: IAEA Advanced Reactors Information System (ARIS), (2020) 187-190, https://aris.iaea.org.

[V–2] IIGAKI, K. and HANAWA, S., Seismic Design, Nuclear Engineering and Design, **233** (2004) 59-70, https://doi.org/10.1016/j.nucengdes.2004.07.011.

[V–3] IIGAKI, K., ONO, M., SHIMAZAKI, Y., TOCHIO, D., SHIMIZU, A., INOI, H., TAKADA, S., and SAWA, K., A Safety Evaluation of HTTR Core Graphite Structures Against the Great East Japan Earthquake, Mechanical Engineering Journal, **1** 4 (2014), https://doi.org/10.1299/mej.2014tep0021.

[V–4] ONO, M., IIGAKI, K., SAWAHATA, H., SHIMAZAKI, Y., SHIMIZU, A., INOI, H., KONDO, T., KOJIMA, K., TAKADA, S., and SAWA, K., Comprehensive Seismic Evaluation of HTTR Against the 2011 Off the Pacific Coast of Tohoku Earthquake, Journal of Nuclear Engineering and Radiation Science, **4** 2 (2018), https://doi.org/10.1115/1.4039288.

[V–5] NUCLEAR REGULATION AUTHORITY, Regulatory Guide on Design Basis Earthquake and Seismic Design Requirements, NRA, Tokyo (2013) (in Japanese), https://www.nsr.go.jp/data/000382458.pdf.

[V–6] ISHIWATARI, A., "Japan's Nuclear Regulation against Natural Hazards after the Fukushima Daiichi Accident", presented at IAEA Technical Meeting on Lessons Learned and Safety Improvements Related to External Hazards Based on the IAEA Fukushima Daiichi Accident Report, Vienna, 2016, https://www.nsr.go.jp/data/000172225.pdf.

[V–7] JAPAN ATOMIC ENERGY AGENCY, Safety measures of HTTR based on new regulatory requirements, JAEA, Tokyo (2021) (in Japanese), https://www.pref.ibaraki.jp/bousaikiki/genshi/anzen/documents/r3_1_siryou1syuusei_1.pdf.

## Annex VI

## EXAMPLE OF SITE INVESTIGATIONS RECOMMENDED FOR A GE HITACHI BWRX-300 REACTOR SITE

This annex provides a brief summary of the practice in an individual State regarding the spacing and depth of boreholes for detailed geotechnical investigation at a nuclear power plant site. The practical implementation of these general guidelines for spacing and depth of boreholes is then illustrated by an example of site investigations carried out for the GE Hitachi BWRX-300 reactor site [VI–1].

### VI–1. GENERAL GUIDELINES FOR SPACING AND DEPTH OF SUBSURFACE INVESTIGATION

Geotechnical investigation and characterization of a site are determined by the foundation requirements of the nuclear installation and the complexity of subsurface materials. In general, the scope of the investigation campaign for a particular site needs to be defined by a team of qualified geotechnical engineers and geologists.

The general guidelines for minimum spacing and depth of subsurface explorations for foundations of structures important to safety, given in Appendix D of Ref. [VI–2], are summarized in Table VI–1. Note that values in Table VI–1 correspond to favourable, uniform geological conditions, where there is horizontal continuity of subsurface strata. The spacing of boreholes needs to be smaller if variable or unfavourable conditions are found.

For favourable and uniform site conditions, as a general guideline, spacing of boreholes is approximately 30 m, with a minimum of three boreholes within the footprint of every structure important to safety.

The minimum depth of boreholes depends on the nature of the ground. For soil sites, the depth of boreholes may be taken as the depth where the vertical stress state will not be altered more than 10% during or after construction by the expected foundation loadings. For footings with rectangular footprints, the minimum depth beneath the foundation plane elevation may be estimated as about twice the smaller dimension of the footprint.

It may be necessary to include several boreholes with penetration depths greater than those needed for other engineering purposes to properly establish the soil model for soil–structure interaction studies to be used in seismic analysis and design. For this purpose, a depth needs to be taken below which (a) the ground will behave elastically for the design level earthquake and (b) deeper sharp impedance[10] contrasts are not to be expected. These two conditions are met if the investigation depth is taken to a strong impedance contrast (e.g. the bedrock), or a horizon with high shear wave velocity $v_s$ (e.g. $v_s > 1800$ m/s).

For medium and hard rock sites, boreholes need to be taken to a greater depth where weak zones, alterations, or discontinuities in the rock massif can affect the foundations. In general, boreholes should be a minimum of 10 m below the lowest elevation of the foundation and should penetrate at least 6 m into the sound rock.

In practice, the geotechnical investigation programme is commonly developed through several iterations. The first iteration is usually exploratory. It has a reduced scope, with the purpose of quick collection of preliminary information about subsurface conditions and horizontal uniformity. In a second iteration, the information collected in the first iteration is used to increase the level of detail. For favourable, uniform geological conditions, the outcome of the second iteration may be sufficient for characterizing the site. If these investigations are not sufficient, the level of detail should be increased with a third iteration of data collection.

Note that planning and implementation of an investigation programme normally involves specialized contractors for field and laboratory testing and for interpretation and consolidation of the results. Planning and implementation of a geotechnical investigations programme in the first iteration can take between six and nine months, whereas the time needed for a second iteration can be between 12 and 18 months.

---

[10] Impedance in this context is the product of mass density and wave speed.

TABLE VI–1. EXAMPLE OF SPACING AND DEPTH OF BOREHOLES FOR GEOTECHNICAL INVESTIGATION FOR FOUNDATIONS OF STRUCTURES IMPORTANT TO SAFETY IN NUCLEAR POWER PLANTS
*(adapted from Ref. [VI–2])*

| Type of structure | Spacing of boreholes | Minimum depth |
|---|---|---|
| General | At least three boreholes within the footprint of every safety-related structure, unless other reliable information is available in the immediate vicinity or otherwise justifiable. | Generally, depth of all boreholes is a minimum 10 m below the lowest part of the foundation.<br>For thick soil sites minimum depth of boreholes is depth $d_{max}$ at which the change in vertical stress for the combined foundation loading is less than 10% of the effective in situ overburden stress. It may be necessary to have a number of deeper boreholes to establish the soil model for soil–structure interaction analyses.<br>For medium/hard rock sites minimum depth of boreholes is 6 m into sound rock (i.e. no discontinuities or zones of weaknesses or alteration which can affect foundations).<br>For weathered shale or soft rock depth is the same as for soils. |
| Buildings | Principal boreholes: One at the centre of the footprint and additional boreholes along the periphery, at corners and other selected locations. At least one borehole per 900 m$^2$ (~30 m spacing) for containment and auxiliary buildings and for linear structures one per 30 m spacing. | At least one-fourth of the principal boreholes and a minimum of one borehole per structure to penetrate into sound rock or to a depth $d_{max}$ at which the change in vertical stress for the combined foundation loading is less than 10% of the effective in situ overburden stress.<br><br>Other boreholes to a depth below foundation elevation equal to the greater of:<br>• Width of the footprint<br>• Foundation depth below the original ground surface |
| Deep cuts, canals | Principal boreholes: one per 60 m along the alignment and at critical locations perpendicular to the alignment to establish geological sections. | Principal boreholes: one per 60 m to penetrate into sound rock or to $d_{max}$. Others to a depth below the bottom elevation of excavation equal to the depth of cut or to below the lowest potential failure zone of the slope. |
| Pipelines | Principal boreholes: One per 30 m, for buried pipelines; at least one for each footing for pipelines above ground. | Principal boreholes: For buried pipelines, one out of every three to penetrate sound rock or to $d_{max}$. Other boreholes, to five times the pipe diameters below the elevation. For pipelines above ground, depths as for foundations for buildings. |
| Tunnels | Principal boreholes: One per 30 m. | Principal boreholes: one per 60 m to penetrate into sound rock or to $d_{max}$.<br>Others to five times the tunnel diameter below the invert elevation. |

Obviously, characterization of a site with the smaller area will need shorter time as compared to a site with large area. In this sense, a site for a small modular reactor (SMR), with a small footprint, will need less time for site characterization than a site for a large nuclear power plant.

## VI–2. AN EXAMPLE OF SITE INVESTIGATIONS RECOMMENDED FOR THE GE HITACHI BWRX-300 REACTOR SITE

The GE Hitachi BWRX-300 reactor is an SMR whose description can be found in the IAEA Advanced Reactor Information System database [VI–3]. Figure VI–1 shows a schematic plant layout. The power block structure, including the reactor building, the control building, the turbine building and the radwaste building, is within a 143 x 69 m rectangular footprint [VI–1].

In the GE Hitachi BWRX-300 SMR design, the majority of the systems and components important to safety are located in a below grade vertical cylinder shaft, within the reactor building. The cylinder is 32.1 m inner diameter. In the standard plant design, the cylinder is embedded about 35 m below plant grade. Figure VI–2 provides a longitudinal cross-section of the power block, including the turbine building, the reactor building and the control building [VI–1].

For licensing of the standard plant design, the subsurface exploration programme shown in Fig. VI–3 has been recommended by the vendor, in accordance with the NRC guidance [VI-2]. This subsurface exploration programme is based on the assumption of favourable and uniform site conditions.

A total of 21 boreholes were planned for a BWRX-300 SMR installation for adequate characterization of subsurface material properties below and surrounding the power block structure. Among those, 16 boreholes were recommended for the power block buildings. Five of them were located within the footprint of the reactor building. Since the footprint of the power block is 143 x 69 = 9867 $m^2$, there is a borehole every 9867/16 = 616 $m^2$. Hence, the average spacing is about 25 m.

One borehole for each of the independent spent fuel storage installation and circulating water pumphouse was planned.

The recommended site investigation plan is summarized in Table VI–2. Keeping in view the embedment depth of the cylinder within the reactor building, three boreholes were taken to a maximum depth of 120 m that correspond to maximum recommended depth for engineering purposes. This depth is greater than both of the following:

— The embedment of the cylinder plus twice the diameter of the cylinder;
— Twice the width of the footprint of the power block buildings.

A total number of seven boreholes were used for geophysical testing, in order to obtain geodynamic properties that are to be used in seismic soil-structure interaction analyses.

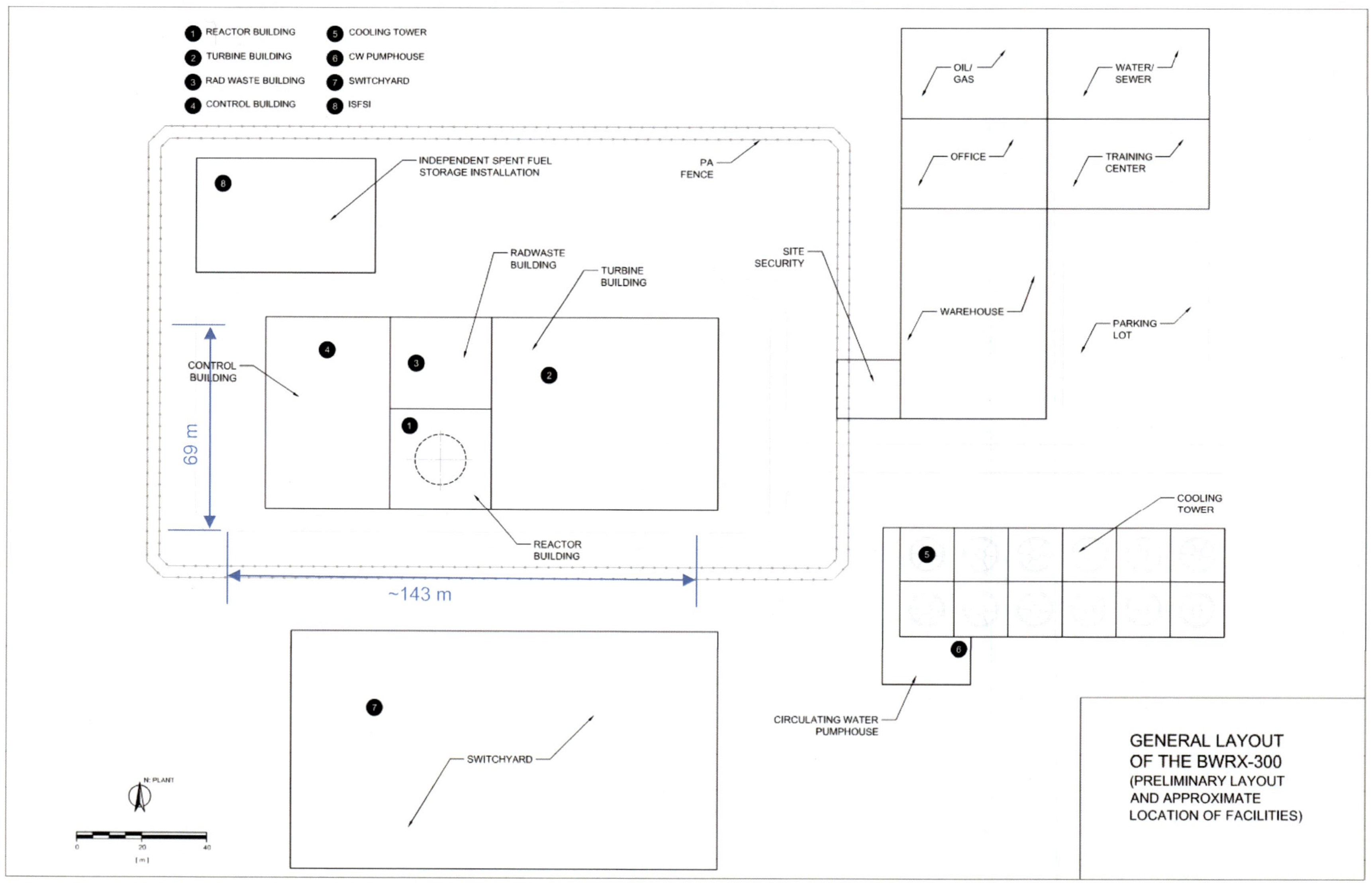


*FIG. VI–1. BWRX-300 plant layout (reproduced from Ref. [VI–1] with permission).*

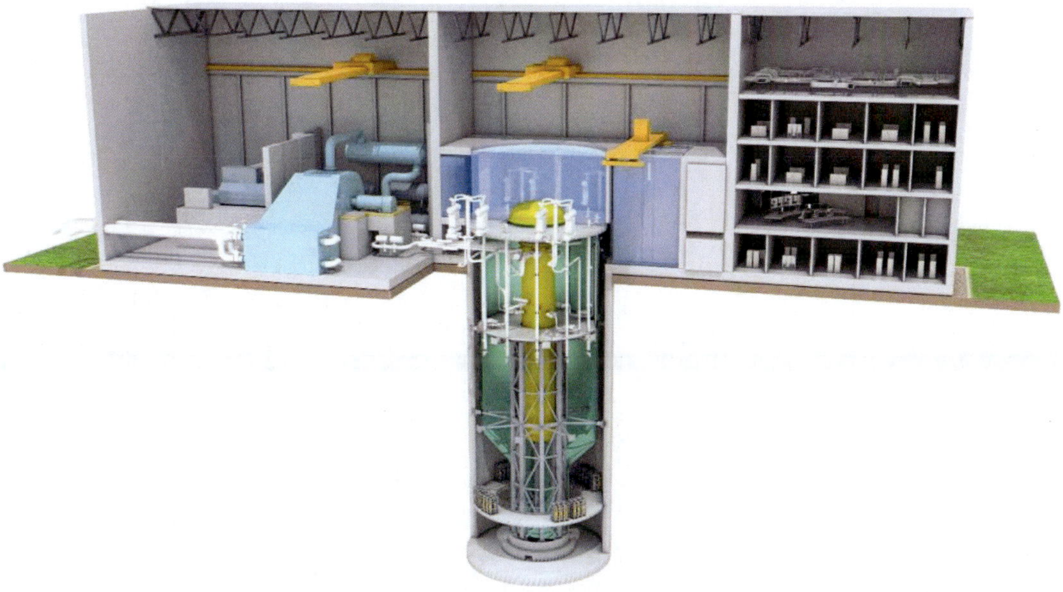

*FIG. VI–2. BWRX-300 reactor building, turbine building and control building cross-sectional view (reproduced from Ref. [VI–3]).*

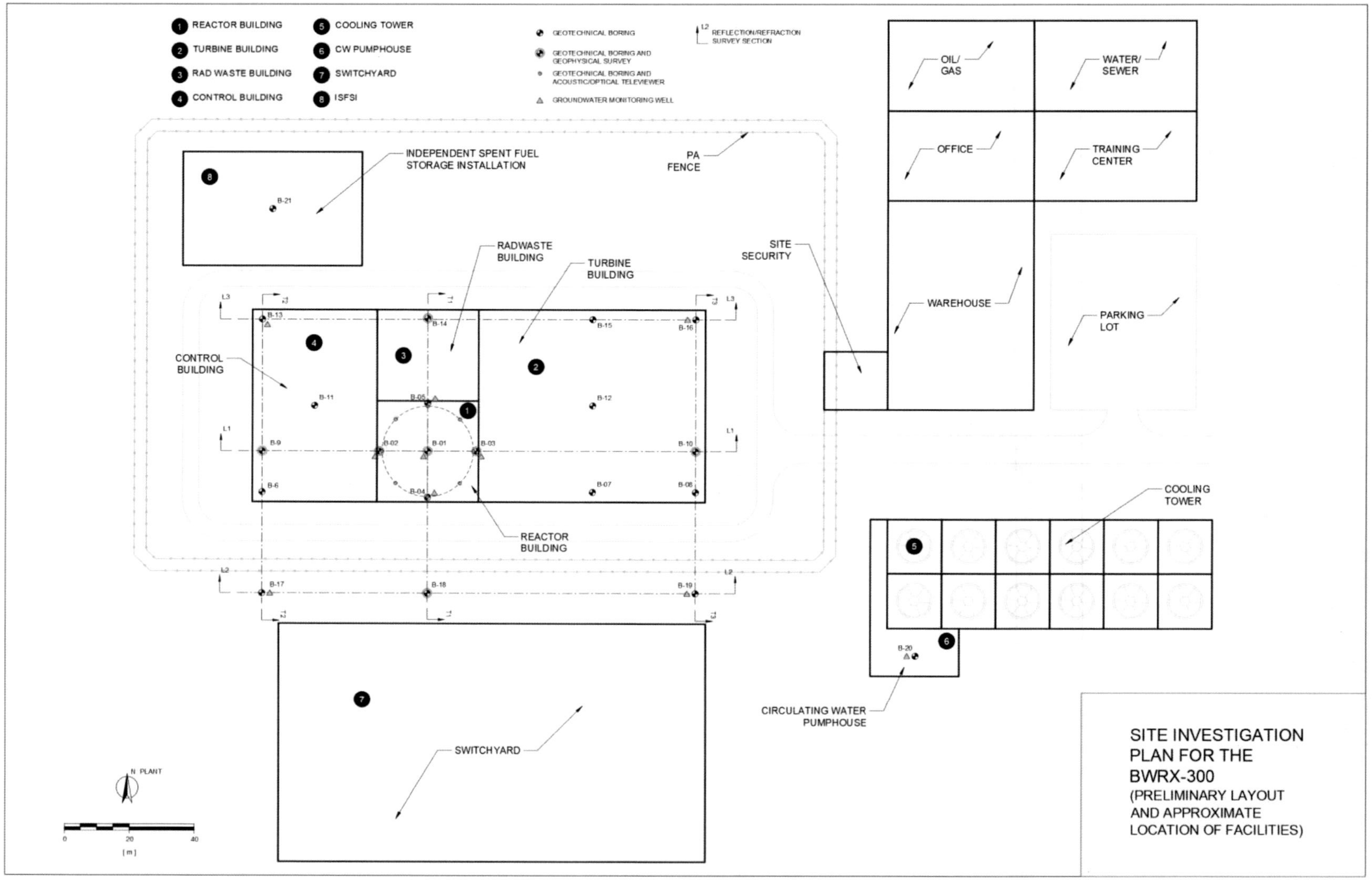


*FIG. VI–3. Preliminary layout of the BWRX-300 boreholes (reproduced from Ref. [VI–1] with permission).*

TABLE VI–2. RECOMMENDED SITE INVESTIGATION FOR THE BWRX-300
*(adapted from [VI–1] with permission)*

| Type of investigation | Purpose | Number |
|---|---|---|
| Geotechnical boreholes | • Measure standard penetration<br>• Measure cone penetration resistance<br>• Sample soils and rock for visual classification and laboratory testing<br>• Rock quality designation<br>• Perform pressure meter tests on weak to moderately soft rock portions to have data parameter for estimation of elastic moduli<br>• Measure in site stress (over coring hydraulic fracturing) | • Three boreholes at perimeter and centre of containment down to 120 m<br>• Two boreholes at perimeter of containment down to a depth of 60 m<br>• About 18 additional boreholes designed to cover the footprint of the main facilities, meet the regulatory guidance and characterize the subsurface as a unit. |
| Wells | • Groundwater characterization (pump and slug tests, baseline groundwater quality)<br>• Characterize groundwater flow direction and quantify hydraulic gradients | • Nine wells at the centre and edge of containment to anticipated depth of 60 m<br>• Four wells down to a depth of 60 m covering the footprint of the facility |
| Geophysical boreholes | • Measure *V*p and *V*s with at least two methods: seismic downhole survey, cross-hole and/or PS logging suspension survey | • One borehole down to 120 m at the centre<br>• Two boreholes at perimeter of containment down to 120 m<br>• Four boreholes located a distance apart from reactor building to allow for wider cross-sections and correlations to refraction or reflection surveys |
| Refraction surveys | • For sites in which a bedrock horizon is identified by the borehole programme, perform seismic refraction to obtain a three dimensional mapping of the bedrock horizon and the thickness of weathered layers | • One grid of surveys covering the footprint extension of the facility |
| Seismic reflection survey | • Identify if voids, sinkholes, karst or faults are present beneath the footprint of the facilities | • Three longitudinal and two to three transverse reflection sections |
| Borehole televiewer | • Observe rock surface directly, subsurface lithology and structural features such as fractures, fracture infillings, foliation and bedding planes<br>• Packer water pressure tests in rock<br>• Measure in site stress (over coring hydraulic fracturing) | • Relevant for rock conditions, over which borehole recovery and rock quality designation allow for an open borehole<br>• The proposed eight televiewer locations will support a better characterization of the rock mass and as a substitute for potential inspection limitations due to the construction process |

## References for Annex VI


[VI–1] GE HITACHI NUCLEAR ENERGY, BWRX-300 Advanced Civil Construction and Design Approach, Licensing Topical Report NEDO-33914 (Rev. 0), ADAMS accession number ML21020A137, NRC, Washington, DC (2021).

[VI–2] NUCLEAR REGULATORY COMMISSION, Geologic and Geotechnical Site Characterization Investigations for Nuclear Power Plants, Regulatory Guide 1.132, NRC, Washington, DC (2021).

[VI–3] INTERNATIONAL ATOMIC ENERGY AGENCY, Advanced Reactor Information System (ARIS), database of advanced reactor design descriptions, https://aris.iaea.org/.

## CONTRIBUTORS TO DRAFTING AND REVIEW

| | |
|---|---|
| Awad, R. | Federal Authority for Nuclear Regulation, United Arab Emirates |
| Beltran, F. | Consultant, Spain |
| Contri, P. | International Atomic Energy Agency |
| Fukushima, Y. | International Atomic Energy Agency |
| Gulerce, Z. | International Atomic Energy Agency |
| Guerpinar, A. | Consultant, Türkiye |
| Khouaja, H. | International Atomic Energy Agency |
| Mahmood, H. | Consultant, Pakistan |
| Mahmood, M. | International Atomic Energy Agency |
| Miller, D. | Canadian Nuclear Safety Commission, Canada |
| Monette, P. | Consultant, Belgium |
| Nagasawa, K. | International Atomic Energy Agency |
| Rocher, M. | Institute for Radiation Protection and Nuclear Safety, France |
| Stout, D. | Ultra Safe Nuclear, United States of America |
| Valentini, A. | International Atomic Energy Agency |
| Yamano, H. | Japan Atomic Energy Agency, Japan |

### Technical Meeting

Vienna, Austria: 28 November–02 December 2022

### Consultants Meetings

Virtual: 29–30 June 2021, 02 September 2021, 08 November 2021, 01–02 February 2022, 14–16 June 2022;
Vienna, Austria: 23–25 November 2022

IAEA
International Atomic Energy Agency

# CONTACT IAEA PUBLISHING

Feedback on IAEA publications may be given via the on-line form available at:
www.iaea.org/publications/feedback

This form may also be used to report safety issues or environmental queries concerning IAEA publications.

Alternatively, contact IAEA Publishing:

Publishing Section
International Atomic Energy Agency
Vienna International Centre, PO Box 100, 1400 Vienna, Austria
Telephone: +43 1 2600 22529 or 22530
Email: sales.publications@iaea.org
www.iaea.org/publications

Priced and unpriced IAEA publications may be ordered directly from the IAEA.

## ORDERING LOCALLY

Priced IAEA publications may be purchased from regional distributors and from major local booksellers.

25-03497E